play

play

How It Shapes the Brain,

Opens the Imagination,

and Invigorates the Soul

STUART BROWN, M.D.,

with CHRISTOPHER VAUGHAN

Avery • a member of Penguin Group (USA) Inc. • New York

Published by the Penguin Group
Penguin Group (USA) Inc., 375 Hudson Street, New York, New York 10014, USA • Penguin Group (Canada), 90 Eglinton Avenue East, Suite 700, Toronto, Ontario M4P 2Y3, Canada (a division of Pearson Penguin Canada Inc.) • Penguin Books Ltd, 80 Strand, London WC2R 0RL, England • Penguin Ireland, 25 St Stephen's Green, Dublin 2, Ireland (a division of Penguin Books Ltd) • Penguin Group (Australia), 250 Camberwell Road, Camberwell, Victoria 3124, Australia (a division of Pearson Australia Group Pty Ltd) • Penguin Books India Pvt Ltd, 11 Community Centre, Panchsheel Park, New Delhi–110 017, India • Penguin Group (NZ), 67 Apollo Drive, Rosedale, North Shore 0632, New Zealand (a division of Pearson New Zealand Ltd) • Penguin Books (South Africa) (Pty) Ltd, 24 Sturdee Avenue, Rosebank, Johannesburg 2196, South Africa

Penguin Books Ltd, Registered Offices: 80 Strand, London WC2R 0RL, England

First paperback edition 2010

Page 231 constitutes an extension of this copyright page.

Most Avery books are available at special quantity discounts for bulk purchase for sales promotions, premiums, fund-raising, and educational needs. Special books or book excerpts also can be created to fit specific needs. For details, write Penguin Group (USA) Inc. Special Markets, 375 Hudson Street, New York, NY 10014.

The Library of Congress has catalogued the hardcover edition as follows:

Brown, Stuart.
Play : how it shapes the brain, opens the imagination, and invigorates the soul / Stuart Brown with Christopher Vaughan.
p. cm.
Includes index.
ISBN 978-1-58333-333-4
1. Play—Psychological aspects. 2. Play—Social aspects. 3. Mind and body.
I. Vaughan, Christopher. II. Title.
BF717.B76 2009 2008050463
155—dc22

ISBN 978-1-58333-378-5 (paperback edition)

Printed in the United States of America
29th Printing

BOOK DESIGN BY NICOLE LAROCHE

To my children:

Caren, who brought me the joy of attunement, and from whom I continue to learn.

Colin, whose life is empowered by play and joy.

Barry, who deftly combines compassion and play, inspiring us all.

Lauren, selfless, inventive, and fun, showing that life can be a playground.

And to their mother, Joan, for grounding them in love.

Contents

Part One. why play?

Chapter One. the promise of play 3

Chapter Two. what *is* play, and why do we do it? 15

Chapter Three. we are built for play 47

Part Two. living the playful life

Chapter Four. parenthood is child's play 77

Chapter Five. the opposite of play is not work 123

Chapter Six. playing together 157

Chapter Seven. does play have a dark side? 175

Chapter Eight. a world at play 195

Acknowledgments 219

Index 223

Part One

why play?

Chapter One

the promise of play

After five hours of driving over the tire-melting highways of the Nevada and Utah deserts, I am beat. My yellow Lab, Jake, shares the emotion. He is draped across the backseat, all the air let out of him. The last ten miles of our journey is an unpaved, rattling road up to my cousin Al's ranch, so it is half an hour more before I shut down the engine and the dust cloud that has been following us blankets the car.

Then something miraculous occurs.

I open the door for Jake and he freezes, every sense aquiver. He instantly takes in the whole scene: a bright August day, four acres of pasture, a dozen horses, my cousin Al, his four kids, and two dogs. A light breeze rustles aspen leaves, wafting scents of hay and horses across the Utah ranch. Doggie heaven.

In half a second Jake is flying out the door, a blond blur zipping toward the pasture. He races at full gallop one way and reverses, paws tearing up the dust in a skidding turn, then accelerates to warp speed in the opposite direction. His mouth is agape, the corners pulled back in a canine grin, his tongue lolling out one side.

Jake blasts into the maze of animals without hesitation. I worry about how the horses will react, but they don't shy. In a flicker the horses are jumping and gamboling. It seems that we all—adults, kids, dogs, horses—recognize that Jake is consumed with the joy of play. All of us are caught up in the moment.

Jake initiates a free-for-all game of follow the leader. He darts from horse, to person, to dog, to pony, to person, and back to horse in an outstanding display of speed, athleticism, and pure exuberance. Jake shoulder-checks another dog and sends him flying, but he doesn't lose a bit of speed and the other dog is right back up and into the chase. The children squeal with delight and run after Jake as he does figure eights. The adults are soon whooping and running. Even some observing magpies get caught up in the act, swooping over the melee.

The moment is captivating, gleeful, unexpected, and short-lived. After thirty seconds the horses scatter and the dogs lie down, panting and cooling their bellies in the grass. All of us feel completely exuberant. We catch our breath and laugh. The tension and fatigue of the drive has fallen from my shoulders. The kids are giggling. The rest of the day has a lightness and ease that I hadn't felt for a long time.

On that day, Jake gave a compact demonstration of what years of academic and clinical research has taught me about the power of play. Most obviously, it is intensely pleasurable. It energizes us and enlivens us. It eases our burdens. It renews our natural sense of optimism and opens us up to new possibilities.

Those are all wonderful, admirable, valuable qualities. But that is just the beginning of the story. Neuroscientists, developmental biologists, psychologists, social scientists, and researchers from every point of the scientific compass now know that play is a profound biological process. It has evolved over eons in many animal species to promote survival. It shapes the brain and makes animals smarter and more adaptable. In higher animals, it fosters empathy and makes possible complex social groups. For us, play lies at the core of creativity and innovation.

Of all animal species, humans are the biggest players of all. We are built to play and built through play. When we play, we are engaged in the purest expression of our humanity, the truest expression of our individuality. Is it any wonder that often the times we feel most alive, those that make up our best memories, are moments of play?

That is something that struck me as I was reading obituaries of those who lost their lives on September 11, 2001, stories I began collecting because they were such poignant and gripping portraits. Soon I realized that what people most remembered about those who died were play moments or play activities. The March 31, 2002, edition of *The New York Times*, to take one example, has obituaries with these headlines: "A Spitball-Shooting Executive," "A Frank Zappa Fan," "The Lawn King: A Practical Joker with a Heart," "A Lover of Laughter." What dominated the profiles beneath the headlines were remembrances of play states with loved ones, which were like joyful threads running through their lives, weaving memories and binding them together emotionally.

I HAVE SPENT a career studying play, communicating the science of play to the public, and consulting for Fortune 500 companies on how to incorporate it into business. I have used play therapies to help people who are clinically depressed. I frequently talk with groups of parents who inevitably are concerned and conflicted about what constitutes healthy play for their kids. I have gathered and analyzed thousands of case studies that I call play histories. I have found that remembering what play is all about and making it part of our daily lives are probably the most important factors in being a fulfilled human being. The ability to play is critical not only to being happy, but also to sustaining social relationships and being a creative, innovative person.

If that seems to be a big claim, consider what the world would be like without play. It's not just an absence of games or sports. Life without play is a life without books, without movies, art, music, jokes, dramatic stories. Imagine a world with no flirting, no daydreaming, no comedy, no irony. Such a world would be a pretty grim place to live. In a broad sense, play is what lifts people out of the mundane. I sometimes compare play to oxygen—it's all around us, yet goes mostly unnoticed or unappreciated until it is missing.

But what happens to play in our lives? Nearly every one of us starts out playing quite naturally. As children, we don't need instruction in how to play. We just find what we enjoy and do it. Whatever "rules" there are to play, we learn from our playmates. And from our play we learn how the world works, and how friends interact. By playing, we learn about the mystery and excitement that the world can hold in a tree house, an old tire swing, or a box of crayons.

At some point as we get older, however, we are made to feel guilty for playing. We are told that it is unproductive, a waste of time, even sinful. The play that remains is, like league sports, mostly very organized, rigid, and competitive. We strive to always be productive, and if an activity doesn't teach us a skill, make us money, or get on the boss's good side, then we feel we should not be doing it. Sometimes the sheer demands of daily living seem to rob us of the ability to play.

The skeptics among the audiences I talk to will say, "Well, duh. Of course you will be happy if you play all the time. But for those of us who aren't rich, or retired, or both, there's simply is no time for play." Or they might say that if they truly gave in to the desire to experience the joy of free play, they would never get anything done.

This is not the case. We don't need to play all the time to be fulfilled. The truth is that in most cases, play is a catalyst. The beneficial effects of getting just a little true play can spread through our lives, actually making us more productive and happier in everything we do.

One example of this is Laurel, the CEO of a successful commercial real estate company. During her late twenties, Laurel married and had two children, all while establishing her business. Her relationship with her husband was close and compatible, and she adored her four- and ten-year-olds. She saw herself as blessed and fortunate.

Her days hummed like a turbocharged engine. Up at five, she usually ran four or five miles on odd days and swam and lifted weights on even days. She didn't work weekends and usually had enough steam left for "quality time" with her supportive husband and kids, church, and her closest friends.

She felt that she had a healthy mix of play and work, but when she passed forty she began to dread her schedule. She didn't yet feel a need to quit any of her commitments or ease off, but slowly she realized that though she had fun with her husband and kids and a sense of enthusiasm about her work, she was missing . . . joy.

So Laurel set about finding where it had gone. She remembered back to her earliest joyful memories and realized they centered on horses. As she reconstructed her own play history, she realized that horses had grabbed her from the first time she saw one. As a toddler she loved bouncing on her hobbyhorse. One of her fondest memories was befriending a local backyard horse and secretly riding it at age seven. She would entice the horse to the fence with carrots and coax it to allow her to climb up and ride bareback, completely unbeknownst to the owner or her parents. As dangerous as it was for a seven-year-old to ride this way, it gave Laurel a sense of her own power. Later she started hanging around stables, becoming an accomplished horsewoman and as a young adult competing as a professional rider. She eventually burned out on horse shows and settled into marriage and business.

Yet she now realized she longed "just to ride."

Laurel decided to make this happen. She found a horse to lease and began to ride again. The feelings of joy and exhilaration came back the first time she climbed onto the horse. Now she makes the time to go riding once a week.

What surprises her most since she incorporated the pure play of riding back into her life is how complete and whole she now feels in all other areas of her life. The bloom of "irrational bliss" she experiences in the care of her horse, from riding it regularly, and even occasionally riding again in small local shows, has spilled over into her family and work lives. The little chores of daily living don't seem so difficult anymore.

She is also surprised by the subtle shift in her relationship with her husband. "It's just easier now. I look forward to talking more often now," Laurel says. Before rediscovering her horse-based play, when she approached her husband for a discussion she was defensively anticipating difficulties or thinking of things that needed doing. "It felt more like job-sharing than being a couple."

At some offices, play is becoming increasingly recognized as an important component of success. And I'm not just talking about Ping-Pong tables in the break room. Employees who have engaged in play throughout their lives outside of work and bring that emotion to the office are able to do well at work-related tasks that at first might seem to have no connection at all to play.

An example: Cal Tech's Jet Propulsion Laboratory (JPL) has been the United States' premier aerospace research facility for more than seven decades. The scientists and engineers at JPL have designed and managed major components of every manned and unmanned mission of our time, and have been completely responsible for dreaming up, building, and operating complex projects like the robot vehicles that landed on Mars and explored the planet's surface for years. You might say that JPL invented the Space Age. No matter how big and ambitious the goal, the researchers could always be relied on to say, "We can do that."

But in the late nineties, the lab's management was saying, "JPL, we have a problem." As the lab neared the new century, the group of engineers and scientists who had come on board in the 1960s, those who put men on the moon and built robotic probes to explore the solar system, were retiring in large numbers. And JPL was having a hard time replacing them. Even though JPL hired the top graduates from top engineering schools like MIT, Stanford, and even Cal Tech itself, the new hires were often missing something. They were not very good at certain types of problem solving that are critical to the job. The experienced managers found that the newly minted engineers might excel at grappling with theoretical, mathematical problems at the frontiers of engineering, but they didn't do well with the practical difficulties of taking a complex project from theory to practice. Unlike their elders, the young engineers couldn't spot the key flaw in one of the complex systems they were working on, toss the problem around, break it down, pick it apart, tease out its critical elements, and rearrange them in innovative ways that led to a solution.

Why was JPL hiring the wrong sorts of engineers? The people JPL brought aboard had earned the highest grades at the best schools, but academic excellence was obviously not the most important measure of the graduates' problem-solving skills. Like good engineers, JPL management analyzed the problem and concluded that when hiring they were looking at the wrong data. Those job candidates good at problem solving and those who were not could be sorted, they believed, if they found the right metrics.

Then the head of JPL found Nate Jones. Jones ran a machine shop that specialized in precision racing and Formula One tires, and he had noticed that many of the new kids coming in to work at the shop were also not able to problem solve. Jones and his wife, who is a teacher, wondered what had changed. After questioning the new kids and the older employees, Jones found that those who had worked and played with their hands as they were growing up were able to "see solutions" that those who hadn't worked with their hands could not. Jones wrote an article about what he had found, which is how he came to the attention of JPL management.

The JPL managers went back to look at their own retiring engineers and found a similar pattern. They found that in their youth, their older, problem-solving employees had taken apart clocks to see how they worked, or made soapbox derby racers, or built hi-fi stereos, or fixed appliances. The young engineering school graduates who had also done these things, who had played with their hands, were adept at the kinds of problem solving that management sought. Those who hadn't, generally were not. From that point on, JPL made questions about applicants' youthful projects and play a standard part of job interviews.

What Laurel discovered through experience, the JPL managers discovered through research: there is a kind of magic in play. What might seem like a frivolous or even childish pursuit is ultimately beneficial. It's paradoxical that a little bit of "nonproductive" activity can make one enormously more productive and invigorated in other aspects of life. When an activity speaks to one's deepest truth, as horseback riding did for Laurel, it is a catalyst, enlivening everything else.

Once people understand what play does for them, they can learn to bring a sense of excitement and adventure back to their lives, make work an extension of their play lives, and engage fully with the world.

I don't think it is too much to say that play can save your life. It certainly has salvaged mine. Life without play is a grinding, mechanical existence organized around doing the things necessary for survival. Play is the stick that stirs the drink. It is the basis of all art, games, books, sports, movies, fashion, fun, and wonder—in short, the basis of what we think of as civilization. Play is the vital essence of life. It is what makes life lively.

When people know their core truths and live in accord with what I call their "play personality," the result is always a life of incredible power and grace. British educator Sir Ken Robinson has spoken about finding such power and grace in the life of dancer Gillian Lynne, who was the choreographer for the musicals *Cats* and *Phantom of the Opera*. Robinson interviewed her for a book he is writing, titled *Epiphany*, about how people discover their path in life. Lynne told him about growing up in 1930s Britain, about doing terribly in school because she was always fidgeting and never paid attention to lessons. "I suppose that now people would say she had ADHD, but people didn't know you could have that then," Robinson says wryly. "It wasn't an available diagnosis at the time."

Instead, school officials told Lynne's parents that she was mentally disabled. Lynne and her mother went to see a specialist, who talked to Gillian about school while the girl sat on her hands, trying not to fidget. After twenty minutes, the doctor asked to speak to Lynne's mother alone in the hallway. As they were leaving the office, the doctor flipped on the radio, and when they were shut in the hallway the doctor pointed through the window back into the office. "Look," he said, and directed the mother's attention to Gillian, who had gotten up and started moving to the music as soon as they left. "Mrs. Lynne," said the doctor, "your daughter's not sick, she's a dancer."

The doctor recommended enrolling her daughter in dance school. When Gillian got there she was delighted to find a whole room of people like herself, "people who had to move to think," as Lynne explained it. Lynne went on to become a principal dancer in the Royal Ballet, then founded her own dance company and eventually began working with Andrew Lloyd Webber and other producers.

"Here is a woman who has helped put together some of the most successful musical productions in history, has given pleasure to millions, and is a multimillionaire," Robinson says. Of course if she were a child now, he adds, "someone would probably put her on drugs and tell her to calm down."

Robinson's story about Lynne was really about the strength and beauty of living in accordance with who she is—which for her meant living a life of motion and music. If her parents and teachers tried to make her into an engineer, Lynne would have been unhappy and unsuccessful.

Ultimately, this book is about understanding the role of play and using it to find and express our own core truths. It is about learning to harness a force that has been built into us through millions of years of evolution, a force that allows us to both discover our most essential selves and enlarge our world. We are designed to find fulfillment and creative growth through play.

Chapter Two

what *is* play, and why do we do it?

what is play? i hate to say

What are we talking about when we talk about play? Though I have studied play for decades, I have long resisted giving an absolute definition of play because it is so varied. For one person, dangling hundreds of feet above the ground, held there by only a few callused fingers on a granite cliff face, is ecstasy. For someone else, it is stark terror. Gardening might be wonderful fun for some but a sweaty bore for others.

Another reason I resist defining play is that at its most basic level, play is a very primal activity. It is preconscious and preverbal—it arises out of ancient biological structures that existed before our consciousness or our ability to speak. For example, the natural tussling of sibling kittens just happens. In us, play can also happen without a conscious decision that, okay, I'm going to play now. Like digestion and sleep, play in its most basic form proceeds without a complex intellectual framework.

Finally, I hate to define play because it is a thing of beauty best appreciated by experiencing it. Defining play has always seemed to me like explaining a joke—analyzing it takes the joy out of it.

I was forced out of this stance by Lanny Vincent, a colleague and friend who is an accomplished business consultant. Lanny and I were making a presentation to a group of Hewlett-Packard engineers, and shortly before I spoke, Lanny asked me what definition of play I planned to present.

I adopted my usual academic stance. "I don't really use an absolute definition," I said. "Play is so varied, it's preverbal, preconscious . . ."

Lanny was having none of it. "You can't go out there without a definition. These are engineers. They design machines. They munch on mountains of specs and wash them down with streams of data. If you don't have a definition they will eat you alive."

Lanny's portrayal of engineers as threatening technological Paul Bunyans was an exaggeration, of course, but he was basically right. Engineers are professional skeptics. To them, good things and useful ideas last, like laws of nature. Engineers build on the bedrock of established fact. They usually regard emotional components of a system as too vague to be useful. But play inevitably has an emotion-laden context that is essential for understanding. I could see that without some foundational definition, they were going to see the field of play as very squishy, marshy ground on which to build.

Luckily, from my own scientific training I knew that what I needed was a good chart. Nothing soothes the restive natives of Techland like charts, graphs, and data. With that in mind, I quickly put together a couple slides laying out the properties of play. Here is what I showed them:

PROPERTIES OF PLAY

Apparently purposeless (done for its own sake)
Voluntary
Inherent attraction
Freedom from time
Diminished consciousness of self
Improvisational potential
Continuation desire

What do these mean? As I explained to the engineers, the first quality of play that sets it off from other activities is its **apparent purposelessness**. Play activities don't seem to have any survival value. They don't help in getting money or food. They are not done for their practical value. Play is done for its own sake. That's why some people think of it as a waste of time. It is also **voluntary**—it is not obligatory or required by duty.

Play also has **inherent attraction**. It's fun. It makes you feel good. It provides psychological arousal (that's how behavioral scientists say that something is exciting). It is a cure for boredom.

Play provides **freedom from time**. When we are fully engaged in play, we lose a sense of the passage of time. We also experience **diminished consciousness of self**. We stop worrying about whether we look good or awkward, smart or stupid. We stop thinking about the fact that we are thinking. In imaginative play, we can even be a different *self*. We are fully in the moment, in the zone. We are experiencing what the psychologist Mihaly Csikszentmihalyi calls "flow."

Another hallmark of play is that it has **improvisational potential**. We aren't locked into a rigid way of doing things. We are open to serendipity, to chance. We are willing to include seemingly irrelevant elements into our play. The act of play itself may be outside of "normal" activities. The result is that we stumble upon new behaviors, thoughts, strategies, movements, or ways of being. We see things in a different way and have fresh insights. For example, an artist or engineer at the beach might have new ideas about their work while building a sand castle. A kid playing tea party might come to understand that good manners and social conventions can provide safety and power rather than being something imposed merely to make her feel uncomfortable. Those insights weren't the reason they played, but they arrived as the result of it. You never really know what's going to happen when you play.

Last, play provides a **continuation desire**. We desire to keep doing it, and the pleasure of the experience drives that desire. We find ways to keep it going. If something threatens to stop the fun, we improvise new rules or conditions so that the play doesn't have to end. And when it is over, we want to do it again.

These properties are what make play, for me, the essence of freedom. The things that most tie you down or constrain you—the need to be practical, to follow established rules, to please others, to make good use of time, all wrapped up in a self-conscious guilt—are eliminated. Play is its own reward, its own reason for being.

I also showed the engineers a framework for play devised by Scott Eberle, an intellectual historian of play and vice president for interpretation at the Strong National Museum of Play in Rochester, New York. Eberle feels that most people go through a six-step process as they play. While neither he nor I believe that every player goes through exactly these steps in this order, I think it's useful to think of play in this way. Eberle says that play involves:

Anticipation, waiting with expectation, wondering what will happen, curiosity, a little anxiety, perhaps because there is a slight uncertainty or risk involved (can we hit the baseball and get safely on base?), although the risk cannot be so great that it overwhelms the fun. This leads to . . .

Surprise, the unexpected, a discovery, a new sensation or idea, or shifting perspective. This produces . . .

Pleasure, a good feeling, like the pleasure we feel at the unexpected twist in the punch line of a good joke. Next we have . . .

Understanding, the acquisition of new knowledge, a synthesizing of distinct and separate concepts, an incorporation of ideas that were previously foreign, leading to . . .

Strength, the mastery that comes from constructive experience and understanding, the empowerment of coming through a scary experience unscathed, of knowing more about how the world works. Ultimately, this results in . . .

Poise, grace, contentment, composure, and a sense of balance in life.

Eberle diagrams this as a wheel. Once we reach poise, we are ready to go to a new source of anticipation, starting the ride all over again.

When I flashed these slides on the screen, I could see the engineers relax, as if they had been lost but now caught sight of a familiar landmark. The rest of the talk went very smoothly, and afterward many of them told me that they saw play in a new light.

The Dutch historian Johan Huizinga offers another good definition of play. He describes it as "a free activity standing quite consciously outside 'ordinary' life as being 'not serious' but at the same time absorbing the player intensely and utterly. It is an activity connected with no material interest, and no profit can be gained from it. It proceeds within its own proper boundaries of time and space according to fixed rules and in an orderly manner. It promotes the formation of social groupings which tend to surround themselves with secrecy."

This parallels the definition I use in many ways, although I don't think the "rules" have to be fixed, or that there even have to be rules at all. I do agree that play often promotes social interaction and that it fosters new terminologies and customs that set a group apart, but it doesn't have to promote secrecy. Indeed, one of the hallmarks of play is that *anyone* can do it.

In the end, for me, all of these definitions fall short. I can create a thousand PowerPoint slides chock-full of diagrams, charts, and definitions, but there is no way to really understand play without also remembering the feeling of play. If we leave the emotion of play

out of the science, it's like throwing a dinner party and serving pictures of food. The guests can understand all they care to about how the food looks and hear descriptions of how the food tastes, but until they put actual food in their mouths they won't really appreciate what the meal is all about.

I've sometimes found that just a few slides of kids playing hopscotch, or a cat playing with string, or dogs playing fetch, creates more recognition and understanding than all the statistical analysis in the world.

why do we play?

Hudson seemed to be a very dead dog. That's what musher Brian La Doone thought as he watched a twelve-hundred-pound polar bear quickstep across the snowfield, straight toward the sled dogs that were staked away from his camp. That November, the polar bears in the Canadian far north were hungry. The sea had not yet frozen, denying the bears access to the seals that they hunted from the ice. La Doone spent much of his life in the polar bear's territory, and judging from the appearance of this particular bear he knew it had not eaten in months. With a skull-crushing bite or a swipe of its massive claws, the bear could easily rip open one of his dogs within seconds.

But Hudson had other things on his mind. Hudson was a six-year-old Canadian Eskimo sled dog; one of La Doone's more rambunctious pack members. As the polar bear closed in, Hudson didn't bark or flee. Instead, he wagged his tail and bowed, a classic play signal.

To La Doone's astonishment, the bear responded to the dog's invitation. Bear and sled dog began a playful romp in the snow, both opening their mouths without baring their teeth, with "soft" eye contact and flattened hair instead of raised hackles—all signaling that each was not a threat.

In retrospect, the play signals began, even before the two came together. The bear approached Hudson in a loping way. His movements were curvilinear instead of aggressively straightforward.

When predators stalk, they stare hard at their prey and sprint directly at it. The bear and the dog were exchanging play signals with these sorts of curving movements as the bear approached.

The two wrestled and rolled around so energetically that at one point the bear had to lie down, belly up: a universal sign in the animal kingdom for a time-out. At another point during their romp, the bear paused to envelope Hudson in an affectionate embrace.

After fifteen minutes, the bear wandered away, still hungry but

seemingly sated by this much-needed dose of fun. La Doone couldn't believe what he'd just witnessed, and yet he was even more astonished when the same bear returned the next day around the same time for another round of frolicking with Hudson. By the third day, La Doone's colleagues had heard about this interspecies wrestling match and his campsite was filled with visitors eager to catch a glimpse of the two new best friends. Every night for a week, the polar bear and Hudson met for a playdate. Eventually, the ice on the bay thickened enough for the famished but entertained polar bear to return to his hunting grounds for seal.

What was it in these animals' nature that was strong enough to overcome hunger and survival instincts? How can two species that don't interact peacefully read each other's intentions well enough to roughhouse and play-fight, when any misunderstanding could become deadly? As I began to look at these sorts of questions, I started to see that play is a tremendously powerful force throughout nature. In the end, it is largely responsible for our existence as sentient, intelligent creatures.

understanding the biology of play

As with the polar bear and the Canadian Eskimo sled dog, you can see an impulse to play in humans. My first scientific clue about the biological importance of play came to me while I was a medical student during my pediatrics rotation at Texas Children's Hospital, part of Baylor College of Medicine in Houston. We would get up early to make rounds. It was an unnerving place at dawn, few adults, no sounds except from the sick kids or the regular beeping and humming of machines that kept them alive.

The kids who ended up in the hospital were for the most part really sick. They had congenital disorders, metabolic disorders, or serious infectious diseases like meningitis. One particular kid that I remember was about two years old and had lymphocytic choriomeningitis, a potentially fatal viral infection that could not be treated with antibiotics. We had to sustain him on IVs, support his vital functions—and keep monitoring him with a battery of laboratory tests—hoping that he would get better rather than worse.

Like most kids who are recovering from a serious illness, he didn't respond to much outside stimuli. But one morning as I walked into his room for my morning rounds, I greeted him with "Hi, Ivan," and he returned my hello with a big smile. Then he reached out to me. His smile was a sign that joy had returned to his life and was an invitation for me to join him in that feeling. I smiled back and held his hand. Later the same day, I checked his lab tests. They showed no change. But the next day's test showed signs of improvement.

I was intrigued. All standard medical signs had shown no change, and yet something was going on in Ivan's body. In a way not measurable by medical tests, Ivan had turned the corner that day. And the very first thing to come back to normal was not his blood sugar, heart rate, blood pressure, blood electrolytes, cell counts, or any of the other twenty-five "objective" signs. What came back first was his smile. This was not just relief from discomfort, but a play signal. When anyone smiles at another person, they are reaching out, engaging in a play invitation as clear as a dog's play bow. Ivan's first visible sign of returning health was an invitation to play.

I noted this surprising fact, but began to understand it only in retrospect, after I had been studying play for some time.

In the years that followed, I studied a range of people from all walks of life—from murderers to businesspeople, socialites, scientists, artists, and even Nobel Prize winners—and systematically mapped how their unique "play histories," a careful review of the role of play in childhood and adulthood, affected their life course. On one end of the spectrum, I studied murderers in Texas prisons and found that the absence of play in their childhood was as important as any other single factor in predicting their crimes. On the other end, I also documented abused kids at risk for antisocial behavior whose predilection for violence was diminished through play.

play in the animal kingdom

By the 1990s, I had studied play and its lack extensively in humans, but I began to realize that if I really wanted to understand what play does for us, I would have to know about how it operates in other animal species. I would have to place the behavior in a biological and evolutionary context. I sometimes say that I'm like James Michener, who begins his book *Hawaii* with lava rising up out of the seabed millions of years ago and ends with hula at the hotel. I needed to look at the really big picture to bring the details into focus.

Interestingly enough, at that point in time, people who had been studying play in humans didn't generally talk to the people who studied it in nonhuman animals, even though there had to be commonalities. I wanted to pull together the human and nonhuman research that was needed to better ground a science of play in evolutionary biology. I found a remarkable expert in animal play behavior, a maverick scholar named Bob Fagen. Fagen had meticulously compiled the world's knowledge of animal play, from aardvarks to Zonotrichia (sparrows). With his background in ethology, mathematical statistics, and biology, he was the world's foremost expert on the nature of animal play and how it had evolved. In addition, he was conducting the world's longest-running observations of animal play in the wild.

I first contacted Bob and his wife, Johanna, in 1989, looking for some of the answers about what play in animals actually is. Which is why, by the summer of 1992, with the support of the National Geographic Society, I found myself thirty feet up in an old-growth cypress with Fagen at his study site on Alaska's Admiralty Island. For ten years, Bob and Johanna deployed video cameras, a Questar spotting scope, computer programs, and more to conduct an intimate surveillance of the grizzly bears on the island. In doing so, they were compiling the longest and most intricate study of animal play in the wild.

I felt fortunate to be learning about animal play from the Fagens, and they had acquainted me with about twenty-eight of the individual bears that frequent Pack Creek. Bob's meticulous observations have granted him worldwide stature in scientific circles.

Bob nudged me and pointed across the tidal flats toward the outlet of the creek, where it flowed into the inside passage of Seymour Canal. We were about an hour's light-plane flight southwest of Juneau, in a pristine wilderness. The feeding bears we had been watching over those two weeks were round-bellied and high-spirited. The salmon were at the peak of their run, and the creek outlet was gold- and silver-tinged with the pulsating bodies of chum and pinks thrashing upstream.

Two juvenile brown (grizzly) bears in the distance were approaching each other across the meadow that abuts the tidal flats. Ears slightly back, eyes widened, mouths open, they began a playful wrestling match that proceeded over several minutes and across the whole field. The two bears went in and out of the rapids, splashed through clear sparkling pools, circled, pirouetted, then stood and leaned against each other, embracing in an upright dance. Periodically they paused, looked at the water, and then, as if under the influence of a master conductor, set at each other mouth-to-mouth, head-to-head, body-to-body, paw-to-paw, in an agile display of bear play. It is as if they had inhaled some cosmic mist filled with joy and were intoxicated by it.

Fully aware of his encyclopedic knowledge of animal play, but filled with the spirit of the unfettered joyful moment we had just observed, I asked, "Bob, why do these bears play?"

After some hesitation, without looking up, he said, "Because it's *fun.*"

"No, Bob, I mean from a scientific point of view, why do they play?"

"Why do they play? Why do birds sing, people dance—for the . . . *pleasure* of it."

"Bob, you have degrees from Harvard and MIT, and an in-depth knowledge of bears. You're a student of evolution, you've written *the* definitive work on all mammals at play—I know you have more opinions about this. Tell me, why do animals play?"

After a long, tolerant silence, during which I felt as if he were a sensitive artist having to explain a sublime painting to a tasteless dolt, Bob relented. He answered reluctantly: "*In a world continuously presenting unique challenges and ambiguity, play prepares these bears for an evolving planet.*"

Like Bob and many other play researchers, I would prefer to look at the ways in which play makes life beautiful, joyful, and fun, rather than look at the nuts and bolts of play's utility. We would rather study the bird of paradise in flight, in the wild, rather than shooting it down to dissect it. One of the wonderful things about play, one of the elements that *makes* it play, is its apparent purposelessness. But does play really have no purpose? The reason I was in Alaska with Bob is that I was surveying what naturalists and animal behaviorists know about the role of play in the animal kingdom. What Bob was saying was that he, too, hated to saddle play with purpose, but after long study and reflection, there did seem to be purpose after all.

Play is incredibly pervasive in the animal kingdom. Examples of the kind of play-fighting that Fagan and I observed in the bears are common, especially in social mammals and smart birds. Among leopards, wolves, hyenas, rats, cats, and dogs, tussling is simply part of growing up. But there are also a number of animals that seem to play well into adulthood. Adult ravens have been observed sliding down a snowy slope on their backs, flying back to the top and sliding down again. Bison will repeatedly run onto a frozen lake and slide on all fours while trumpeting exultantly. Hippos in the water will do backflips over and over again.

Other researchers and I used to think that play was found only in mammals, birds, and some reptiles, not lower orders. But animal-play researchers have established specific criteria that define play behavior, and it seems that the farther down the evolutionary ladder they look, they still find it. Octopuses, which have developed along an ancient evolutionary line far removed from us, are one of the most studied creatures in the neurosciences. When animal behaviorists observe octopi engaged in "relaxed, idiosyncratic manipulation of objects," especially when it seems to be a kind of stimulus-seeking behavior, they have little choice but to say this satisfies the definition of play. Certain territorial fishes engage in bubble blowing that appears to be play. The esteemed ant expert E. O. Wilson feels that ants engage in play-fighting. Now I see play where I never imagined it would be.

play with a purpose

Again, one of the hallmarks of play is that it appears purposeless. But the pervasiveness of play throughout nature argues that the activity must have some purpose after all. Animals don't have much leeway for wasteful behaviors. Most live in demanding environments in which they have to compete to find food, compete with other species, and compete to mate successfully. Why would they waste time and energy in nonproductive activity like play? Sometimes play activity is even dangerous. Mountain goats bound playfully along rock faces thousands of feet high, and sometimes they fall. As a mountain goat mother might say, "It's all fun and games until someone gets hurt."

As a scientist, I know that a behavior this pervasive throughout human culture and across the evolutionary spectrum most likely has a survival value. Otherwise, it would have been eliminated through natural selection. All else being equal, the mountain goats that are not inclined to play would survive better (they wouldn't fall off the cliff doing some unnecessary stunt) and would pass on their genes more successfully. Over time, if play had no benefit, the playful goats would be pushed out of the gene pool by the offspring of the nonplayers. But that is not what happens, so there must be some advantage to play that offsets the greater risk of death in playful goats.

In fact, play can be scientifically proven to be useful. After carefully documenting the play behavior of the Alaskan grizzlies over more than fifteen years, the Fagens analyzed the results and were able to differentiate play from all other behaviors (the observational criteria and statistical analysis are not easy to summarize, but they are quite specific and constitute statistically significant data). They found that the bears that played the most were the ones who survived best. This is true despite the fact that playing takes away time, attention, and energy from activities like eating, which seem at first glance to contribute more to the bears' survival.

The real question, then, is why and how play is useful. One major theory is that play is simply practice for skills needed in the future. The idea is that when animals play-fight, they are practicing to fight or hunt for real later on. But it turns out that cats that are deprived of play-fighting can hunt just fine. What they can't do—what they never learn to do—is to socialize successfully. Cats and other social mammals such as rats will, if seriously missing out on play, have an inability to clearly delineate friend from foe, miscue on social signaling, and either act excessively aggressive or retreat and not engage in more normal social patterns. In the give-and-take of mock combat, the cats are learning what Daniel Goleman calls emotional intelligence—the ability to perceive others' emotional state, and to adopt an appropriate response.

"I believe that play teaches young animals to make sound judgments," Bob Fagen told me that day in Alaska. "For instance, play-fighting may let a bear learn when it can trust another bear and, if things get too violent, when it needs to defend itself or flee. Play allows 'pretend' rehearsal for the challenges and ambiguities of life, a rehearsal in which life and death are not at stake."

Play lets animals learn about their environment and the rules of engagement with friend and foe. Playful interaction allows a penalty-free rehearsal of the normal give-and-take necessary in social groups. In the animal world, it is common to see a kitten, puppy, or cub playfully lunge and bite at its mother. This pouncing practice may serve them well later in a fight or hunt, but the more important lesson may be how to show off for siblings or learn just how much abuse Mom can take before she freaks out.

In humans, verbal jousting may take the place of physical rough-and-tumble play. Kids at play can learn the difference between friendly teasing and mean-spirited taunting as they explore the boundaries between those two, and learn how to make up when the boundary is crossed. Adults at cocktail parties learn similar social guidelines about how to get along with others, or how to seem to.

the brain on play

Animals that play a lot quickly learn how to navigate their world and adapt to it. In short, they are smarter. Neuroscientist Sergio Pellis of the University of Lethbridge in Canada, and neuroscientist Andrew Iwaniuk and biologist John Nelson of Monash University in Melbourne, Australia, reported that there is a strong positive link between brain size and playfulness for mammals in general. For their study, which was the most extensive quantitative comparative study of juvenile play ever published, they measured brain size and tabulated play behavior in fifteen species of mammals that ranged from dogs to dolphins. They found that when they made allowances for differing body size, the species with larger brains (compared with body size) played a lot and the species with smaller brains played less.

Another renowned senior play researcher, Jaak Panksepp, has shown that active play selectively stimulates brain-derived neurotrophic factor (which stimulates nerve growth) in the amygdala (where emotions get processed) and the dorsolateral prefrontal cortex (where executive decisions are processed).

John Byers, an animal play scholar interested in the evolution of play behavior, has undertaken a detailed analysis of brain size correlated with the degree of playfulness and the relative rung of the evolutionary ladder to which the player belongs. He discovered something: the amount of play is correlated to the development of the brain's frontal cortex, which is the important brain region responsible for much of what we call cognition: discriminating relevant from irrelevant information, monitoring and organizing our own thoughts and feelings, and planning for the future. In addition, the period of maximum play in each species is tied to the rate and size of growth of the cerebellum. This part of the brain lies in back of and below the main hemispheres, and contains more neurons than the whole rest of the brain. Its functions and connections were once thought to be primarily for coordination and motor control, but through new brain-imaging techniques researchers are finding that the cerebellum is responsible for key cognitive functions such as attention, language processing, sensing musical rhythm, and more.

Byers speculates that during play, the brain is making sense of itself through simulation and testing. Play activity is actually helping sculpt the brain. In play, most of the time we are able to try out things without threatening our physical or emotional well-being. We are safe precisely because we are just playing.

For humans, creating such simulations of life may be play's most valuable benefit. In play we can imagine and experience situations we have never encountered before and learn from them. We can create possibilities that have never existed but may in the future. We make new cognitive connections that find their way into our everyday lives. We can learn lessons and skills without being directly at risk.

So how do we create these "simulations"? Through watching and engaging in sports, physical activities, books, storytelling, art, movies, and much, much more. By living through Rick and Ilsa's doomed romance in *Casablanca*, we learn a little bit about love and how to live our lives with honor and a sense of irony when love is lost. When we really get into following the victories and defeats of a favorite football team, we learn about perseverance and how to argue with our friends (about who is the best quarterback, for instance) in a constructive way. When we experience a new physical challenge like learning to ski, we may find that the things we learn on the slopes—like avoiding falling by keeping our weight forward and committing to the turn—may come to mind during business negotiations as important reminders to press forward and commit to the deal—or fail.

On the basis of highly technical research and his speculations stemming from it, the Nobel laureate and neural scientist Gerald Edelman has created a theory about how new information is functionally integrated into the brain. When I correlate his opinions with my observations on how play may craft the developing brain, what he says makes good sense to me. Edelman describes how our perceptual experiences are coded within the brain in scattered "maps," each of which is a complex network of interconnected neurons. For instance, the many different shapes and sizes of trees that exist in the world are encoded into a common map that encodes what "treeness" is, allowing us to recognize a tree even when we have never seen its particular kind before. In this way the brain achieves a rich and flexible series of maps that permit the recognition of innumerable sorts of objects, sounds, colors, social settings, and so on.

The perceptual generalizations arising from these maps are not static. They flex and change. They also have emotional connotations.

We find our way in the world by navigating this huge and organically growing cartography of life.

The vitality of these maps depends on the active and incessant orchestration of countless details. It seems likely that this orchestration happens most fully through play. The act of pretend playing, for example, is a rich stew of mixed perceptions. Imagine a three-year-old sitting on the floor, playing with a stuffed animal, talking to it in various voices. This child is forming neural connections that make more and more sense as they are added to the growing body of stored, mapped information. The very rich connections among the brain's maps are reciprocal and may involve millions of fibers. My sense of these interconnecting and dynamic maps is that they are most effectively enriched and shaped by the "states" of play.

Play's process of capturing a pretend narrative and combining it with the reality of one's experience in a playful setting is, at least in childhood, how we develop our major personal understanding of how the world works. We do so initially by imagining possibilities—simulating what might be, and then testing this against what actually is.

Though this may seem to be a primarily childish trait, close examination of adult internal narratives (our stream of consciousness) reveals something similar. Our adult imaginations are also continually active, predicting the future and examining the consequences of our behavior before it takes place. Just as in children, adult streams of consciousness are enriched through the simulations of childlike imaginative play. We all daydream about events in our future—even if we are not consciously aware of it. These thoughts leave an imprint on our brains. Someone might not even notice as they fantasize about what kind of house they would like to live in, or whom they would like to marry, but the brain is constructing a working profile of a future house or future spouse. Psychoanalyst Ethel Person writes that, through therapy, one client discovered that much of his effectiveness in business came from his repeated imaginings of possible interactions that he might have on a particular issue. By the time he actually had the conversation, he was usually pretty well prepared for any contingency.

The genius of play is that, in playing, we create imaginative new cognitive combinations. And in creating those novel combinations, we find what works.

One biologist who studied river otters decided to train some of them to swim through a hoop by offering a food reward for completing the task. Shortly after the otters learned to do this, the animals started introducing their own twists to the task. They swam through the hoop backward and waited to see if they got a reward. They swam through and then turned around and swam back through the other way. They swam halfway through and stopped. After each variation, they waited expectantly to see if this version of the task would earn a reward or not.

Through their behavior, the otters were testing the system. They were learning the rules of the game, the rules that govern their world. This was not a thought-out strategy. Otters are naturally extremely playful and are always attracted to new and interesting things. Their natural search for novelty and avoidance of boredom leads them to try the task a number of different ways. By having fun and mixing it up, the otters were learning far more about the way their world works than if they had simply performed the initial task flawlessly. It's a lesson we all could learn. The biologist ruefully noted that he had been trying for years to get his graduate students to use such playful investigation rather than rote learning and mechanical thinking in their research.

Landmark research done in the 1960s at UC-Berkeley by Marian Diamond also points to the essential role of play in brain development. One warm winter day I went to visit Diamond, a charming and gracious woman who has also been a groundbreaking neuroscientist for nearly a half century. She was uncovering the secrets of neurological development when few other women were top scientists, much less neuroscientists.

Diamond's name isn't known widely outside scientific circles, but her work is familiar to every parent. In the early 1960s, Diamond and her colleagues conducted the landmark experiments showing not only that rats raised in an "enriched" environment became smarter, but their brains were larger and more complex, with a thicker and more developed cortex—the "gray matter" where the brain's real data processing takes place.

The idea quickly took hold in the popular imagination: If babies were raised in an enriched nursery, with lots of colorful murals and mobiles, they would also experience supercharged brain development.

What Diamond told me about her experiments, though, brought to light an important distinction between her work and its interpretation in popular culture. The rats that grew bigger, more complex brains and became smarter weren't just exposed to a greater variety of stimuli. They weren't merely given more colorful surroundings and more interesting sounds. The secret to brain growth for the rats in the original experiments was that they played with an ever-changing variety of rat "toys" and socialized with other rats.

"The combination of toys and friends was established early on as vital to qualifying the environment as 'enriched,'" Diamond said.

Play was the true key for the rats' brain development. They tussled and chewed, wrestled with each other, explored and interacted with the toys; they investigated and invited other rats to play. Those were *active* things they did. The rats were not passively soaking up their interesting surroundings.

For human babies, the lesson should be not so much that babies should be provided with bright, colorful, interesting nurseries (although this can't hurt). The lesson should be that it is crucial to provide babies and young children with the chance to play and socialize—toys and tots, play and parental interaction—to help them reach their full potential.

Merely changing the surroundings or offering varied challenges was not enough to get dramatic brain development, Diamond found. In one series of experiments, rats were tasked with finding their way through various mazes to find a reward. This solitary, nonplay activity resulted in neural growth in only one area of the brain, as opposed to the whole-brain growth that play provided.

I think that part of the confusion on the part of parents and pundits may have arisen from the term "enrichment," which sounds less like a play activity than an ingredient you can add to the child-raising stew, and by the lack of discussion of the play aspects of the experiments. Diamond says she still finds the term "enrichment" fitting for what they were doing, but she acknowledges that she avoided discussing "play" or "toys" when describing the experiments.

"Back in the early 1960s, women had to struggle to be taken seriously as scientists," Diamond said. "I was already seen as this silly woman who watched rats play, so I did avoid the words 'toys' and 'play.' "

Diamond's experiments are merely among the most well-established research findings showing that play is crucial to healthy brain development. What is the link between neural growth and play? Why do play activities seem to go hand in hand with brain development? What difference does play make? The truth is that play seems to be one of the most advanced methods nature has invented to allow a complex brain to create itself.

Why do I say this? Consider the fact that there is no exact blueprint for creating the brain. The information encoded in our DNA is far too sparse to define exactly how all the neurons should connect up with each other. Instead, the brain wires itself up. It does this by creating far too many neurons, which in turn make far too many connections with other neurons throughout the brain. Following rules of interaction laid down in the DNA, the neurons send signals through the circuits, strengthening those that work and weakening or eliminating those that don't.

This process continues throughout life, and is a kind of neural evolution. After birth, most neurons are already in place, but they continue to make new connections. The fittest connections, the ones that work best, are the ones that survive. It's survival of the fittest.

REM sleep, or dreaming sleep, seems to be a critical part of this testing. Sleep and dreams appear to be organizers of higher brain function. While no one is certain yet about all the functions of sleep and dreams, researchers find that these activities seem to create a dynamic stabilization of the brain and improve memory throughout life. Studies have shown that people remember things better if they have a good night of sleep after learning something. We know that REM sleep is most frequent during the periods of most rapid brain development, and the theory is that, during development, sleep and dreams probably contribute to this testing and strengthening of brain circuits.

Play, which is more prevalent during the periods of most rapid brain development after birth (childhood), seems to continue the process of neural evolution, taking it even one step farther. Play also promotes the creation of new connections that didn't exist before, new connections between neurons and between disparate brain centers. It is activated from and organizes what I call "divinely superfluous neurons." These are neural connections that don't seem to have an immediate function but when fired up by play are, in fact, essential to continued brain organization.

In playing we foster the creation of those new circuits and test them by running signals through them. Because play is a nonessential activity, this testing is done safely, when survival is not at stake. Play seems to be a driving force helping to sculpt how the brain continues to grow and develop.

In rats, at least, the same areas of the brain stem that initiate sleep initiate play behavior. Like sleep, play seems to dynamically stabilize body and social development in kids as well as sustain these qualities in adults. I find it exciting to see parallels between these two major behaviors—sleep and play. It's reasonable to see them both as essential long-term organizers of brain development and adaptability.

the drive to play

Play seems to be so important to our development and survival that the impulse to play has become a biological drive. Like our desires for food, sleep, or sex, the impulse to play is internally generated.

All drives are not equal in strength. Our primary need is to survive from one day to the next. The strongest drives are for food and sleep. When we are in peril, play will disappear. But studies show that if they are well fed, safe, and rested, all mammals will play spontaneously.

As the philosopher Jeremy Bentham observed, our behavior is determined largely by pleasure or pain. We are rewarded for behavior that conforms to the dictates of the biological drives and punished for behavior that goes against them. We feel pain when we don't eat, and great pleasure when we are finally able to chow down (as the saying goes, "Hunger is the best sauce"). A great night's sleep, especially after a string of sleepless nights, is one of the most satisfying, free pleasures available.

As children, our reward for play is strong because we need it to help generate a rapidly developing brain. As adults, the brain is not developing as rapidly and the play drive may not be as strong, so we can do well enough without play in the short term. Our work or other responsibilities often demand we set play aside. But when play is denied over the long term, our mood darkens. We lose our sense of optimism and we become anhedonic, or incapable of feeling sustained pleasure.

There is laboratory evidence that there is a play deficit much like the well-documented sleep deficit. And just as a sleep deficit generates a need for extra "rebound" sleep to catch up, laboratory research shows that animals that are deprived of play will engage in "rebound" play when allowed to do so again. While we don't have statistical evidence that the same happens in humans, anecdotal evidence from parents and teachers, as well as data gathered in many adult play histories I've conducted, indicate that humans also feel a much more intense desire to play when they have gone a long time without it.

The flip side of the play drive is what it does for us when engaged. From the same play histories, I believe that we have anecdotal evidence that with enough play, the brain works better. We feel more optimistic and more creative. We revel in novelties—a new fashion, new car, a new joke. And through our embrace of the new we are attracted to situations that test skills we do not need now, but may need in the future. We find ourselves saying, "I did it just for the heck of it, but it turned out to be good for me."

In an unpredictable, changing world, what we learn from playing can be transferred into other novel contexts. We seek out a variety of new contingencies through play, allowing us to thrive anywhere in the world. The first steam engine was a toy. So were the first airplanes. Darwin got curious about evolution initially through collecting samples from the seaside and garden where he played as a kid. Throwing stones likely led to the first projectiles, and perhaps the first spear. Fireworks in China preceded the cannon. As I muse on this, I think that math likely came via play with numbers. Wind-up toys led to the development of clocks.

When we are not up against life or death, trial and error brings out new stuff. We want to do this stuff not because we think that paper airplanes will lead to 747s. We do it because it's fun. And many years later, the 747 is born.

is the universe playful?

I like to say that when you open your eyes to it, play is everywhere. And I mean that literally—play may operate at all levels, from the smallest cellular interaction to the far reaches of the universe.

Play can be seen as a key component of evolution itself. The part of evolution that gets the most attention is natural selection, which is often called the "survival of the fittest." But there is another part of the process that is equally important: the generation of diversity. First nature generates many different versions of organisms, mostly through gene mutation and gene recombination, and then the best are "selected" by nature to reproduce and pass on their genes. The creation of these oddities, which Darwin called "sports," is a kind of play. They are nonessential creations outside of everyday norms. Their creation adds a flexibility to the biological system. Biologists have shown that when this genetic flexibility is large, evolution proceeds more quickly. If this variation is absent, evolution will cease. Nothing changes.

Indeed, this sort of flexibility or play seems to be an essential part of any complex, self-organizing system. Without odd variations thrown in, systems proceed in lockstep fashion. On a cosmic scale, the formation of galaxies, stars, and solar systems was possible because of slight irregularities in the fabric of the universe that came into existence shortly after the Big Bang. Without these irregularities, the universe would be a homogenous soup of energy. Play is the swing off the rhythm in music, the bounce in the ball, the dance that delivers us from the lockstep march of life. It is the "meaningless" moment that makes the day memorable and worthwhile. I believe we live in a playful universe.

Though my sense of this comes from cosmology and biology, the Hindu tradition formalizes play as the ultimate creative source of reality. *Lila* (Sanskrit) is a concept meaning "pastime," "sport," or "play." *Lila* is a way of describing all reality, including the cosmos, as the outcome of creative play by the divine absolute.

Chapter Three

we are built for play

The sea squirt is an ugly creature. In its adult form it has a tubular shape that resembles a sponge or worm, and in its larval form it looks like a tadpole. Still, the sea squirt is one of our most ancient relatives. Its primitive nervous system makes it more closely related to humans than the sponges and corals it resembles. Scientists say a sea squirt tadpole approximates what an early human ancestor—the very first chordate—may have looked like some 550 million years ago. In this larval form, it has a primitive spinal cord and bundle of ganglia that act as a functional brain. This tiny brain helps it move selectively toward nutrients and away from harm. Like most oceanic creatures, juvenile sea squirts spend their time growing and exploring the sea.

Once the sea squirt grows to adulthood, it attaches itself permanently to a rock or a boat's hull or pilings. It no longer needs to monitor the world as it did as a juvenile because the passing current provides enough nutrients for it to survive. Its life becomes purely passive.

The adult sea squirt becomes the couch potato of the sea. In a surprisingly macabre twist, the sea squirt digests its own brain. Without a need to explore or find its sustenance, the creature devours its own cerebral ganglia. It's like something out of a Stephen King book: "All work and no play make sea squirt a brain-eating zombie."

The sea squirt is an example of a basic principle of nature: Use it or lose it. If a capability is not being used, it becomes an extravagance that is jettisoned or fades away. Either we grow and develop or we waste away.

Most animals don't go to extreme measures like the sea squirt, but the pattern remains the same. Most animals grow new nerve connections extensively only during the juvenile period. The sea squirt stops moving, and many higher animals stop playing, and the brain stops growing.

But not humans. The brain can keep developing long after we leave adolescence and play promotes that growth. We are designed to be lifelong players, built to benefit from play at any age. The human animal is shaped by evolution to be the most flexible of all animals: as we play, we continue to change and adapt into old age. Understanding why many animals stop playing in adulthood, and why humans don't, helps us further understand the role play has in adult life.

playing the hand that's dealt

If play is so good, why do animals ever stop? I've shown how play is an essential part of the development process, which is why all young animals play a great deal. Play creates new neural connections and tests them. It creates an arena for social interaction and learning. It creates a low-risk format for finding and developing innate skills and talents.

The fact is that play is not completely without cost. Play can be dangerous. Australian scientist Robert Harcourt showed in a study of seal pups that among the twenty-six that were killed by predators, twenty-two were killed while playing out of the protective range of their parents. While engaged in play, animals are not finding food or shelter. If adult animals do nothing but play, they won't be paying attention to their offspring, making them more vulnerable to predators. As you may recall, the fact that play has real costs is one of the main reasons that I knew that it must be important. If play is so pervasive in the animal world, despite its costs to the organism, there must be a very strong reason for its existence. There must be a benefit that is even greater than the cost.

The great benefits of play, as I've said, are the ability to become smarter, to learn more about the world than genes alone could ever teach, to adapt to a changing world. These benefits are most effective as the brain is growing most rapidly, during the juvenile period. Once this period ends and development slows, for some animals the costs begin to outweigh the benefits.

In life's poker game, animals are dealt their genetic hand, they exchange a few cards during development, and then it's time to play their hand and see if they win or lose. This basic reproductive game can work well for organisms that reproduce once and die. They grow and learn, but at some point school's out and it's time to see if those learned skills allow them to survive and pass along their genes.

This is the strategy of the salmon that hatch in streams, mature in the ocean, and then get one shot at surviving the merciless upriver marathon to spawn in the gravel streambeds where they were born. It's a primitive pattern usually associated with animals that have many offspring and rely on quantity rather than quality to have the best chance of reproductive success. Fish lay hundreds of eggs and leave, letting their offspring grow on their own and hoping that a few among these hundreds of newborns will grow to maturity.

What is the best strategy, though, for organisms that reproduce several times during their life cycle and who have relatively few offspring during each reproductive cycle? Mammals—humans included—and birds, for instance, have few offspring and need to stick around to protect them in order to provide the best chance that some will survive to reproduce in turn. They also need to have multiple reproductive opportunities so that if the first batch of young ones doesn't survive, they will have another chance (this was a much more significant problem for humans in past history, before we were able to reduce our high infant mortality rate). They need to keep learning and growing, even when they reach their reproductive period, so that they have the chance to fail, learn, and then succeed. For these animals, nature alters the developmental program to allow extra playtime.

Along with our opposable thumbs and massive prefrontal cortex, a singular characteristic of humans is that we stretch our juvenile period out longer than any other creature. Since one of the primary hallmarks of being juvenile is the desire and capacity to play, what would happen if our brains really keep juvenile elements such as growth and adaptability long past the period of our obvious prolonged childhoods? What if the maintenance of very useful juvenile qualities in the brain is the secret to success in many species—especially ours?

the labrador and the wolf

The first time I came face-to-face with a wolf I had the shocking realization that this was not a dog. Of course, that is something we all know logically, and yet they look so similar that we tend to think of them as being the same. I got to see wolves up close when I spent a week with C. J. Rogers, a researcher in New Mexico whose Ph.D. in animal behavior I had helped supervise. I had concluded that she is to wolf studies what Jane Goodall is to the study of chimpanzees.

Dogs, which have been selectively bred over millennia to interact with humans, see us as their pack masters, depend on us for sustenance, and gladly transfer their affections to us. So they greet us with a doggie version of a smile, they wag their tails, and often signal their readiness to play with a play bow. Even as they bark or whine at us, there is a comfortable sense of familiarity to this attention-getting display. In contrast to the complex "singing" in wolves, dogs sound off in response to territorial invasions or frustration, and as an attention-getting or emotional expression. Their play and their aggression are easily communicated to us.

Wolves, even those that have been brought up around humans, are very different. They really don't need us for their survival, and thus don't spontaneously invite human-wolf play. Their regard for humans seems blank by comparison, neither friendly nor angry. Their social structure is complicated and hierarchical, and unless we

participate as a member of the pack (as C.J. does so skillfully) we are intruders. Yet as pups, wolves and dogs are like rambunctious cousins. Wolf pups and canine puppies behave so similarly that they seem virtually identical, distinguishable only by the various features we associate with a particular breed. Each starts life with a pug-nosed muzzle, floppy ears, and strong attachment to their mothers. Wolves pass through predictable stages of development which are characterized by particular kinds of behaviors. As they grow, wolf pups and domestic dogs show an eagerness to play. For a brief period of development, wolf pups are compulsive retrievers. While a wolf pup is in this stage, its muzzle and ears are similar to those of an adult Labrador or golden retriever.

If the domestic puppy is a Lab or golden retriever, its physiological and social development beyond this stage essentially stops. But for a wolf this is just one stage in the path toward wolf adulthood.

The wolves become "pointers" and finally, as they mature, their muzzle is elongated and pointed and their ears stand briskly upright; they have achieved wolf adulthood. At the same age, the Lab is certainly an adult Lab, fully capable of reproduction and no longer puppy size, but behaviorally still acts like a play-and-retriever-stage wolf puppy. Certain breeds of dogs (German shepherds, huskies, poodles) are more wolflike than Labs—they are more loyal and territorial. But golden retrievers and Labs die of old age primarily still players and retrievers.

Yes, wolves can still be playful as adults, but much of the time they are engaged in the business of pack formation and working out their specific status level. Once acquired, this role is much more fixed than the fluid relationships that result from domestic dog socialization. It is geared toward group survival by forming a rigid, hierarchical, but cooperative pack able to function as carnivores in the wild.

Dogs are displaying an adaptive pattern called neoteny (from the Greek for "stretch" or "extend"), which describes the stretching of juvenile periods and sometimes the retention of juvenile characteristics into adulthood. This is a major theme in evolution. Since early development is a time when the nervous system is most "plastic," an advantage that neoteny bestows is extended openness to change, and sustained curiosity, as well as the ability to readily incorporate new information. A seasoned alpha wolf may be a premier hunter but will inevitably remain bound by narrower and more compulsive behaviors than a domestic dog.

Like retrievers, humans are the youthful primates. We are the Labradors of the primate world. Just as Labrador and wolf pups look and act alike, chimpanzee babies look very much like human babies, with high, rounded foreheads and big eyes. As chimpanzees grow older, however, they acquire a sloped forehead and heavy brow ridges, a jaw that juts forward, and look very different than when they were young. They sort of resemble our Neanderthal ancestors. We modern humans remain baby faces all our lives, never losing the high forehead, rounded skull, and other characteristics typical of youth. We look more like chimpanzee babies than the chimp parents.

Not only do we look more like chimpanzee infants than adults, we act more like them. As with the adult wolf, the adult chimp exhibits more compulsive, rigid, and purpose-driven behavior. Adult male chimpanzees have a strict dominance hierarchy, don't play very much unless cajoled by a juvenile to participate, are reactive to strangers approaching their territory, and seem to like to fight more

than play. Baby chimps exhibit the kind of playfulness that looks more human.

This quality of retained "immaturity" goes deeper than our round faces and essentially hairless bodies. The nervous system of adult chimps, if damaged, has less room for repair. We, on the other hand, have much more capacity for new neuron growth, a characteristic of being forever young. A stroke patient in a modern rehabilitation center demonstrates that if the damage is not too severe or localized in a critical area, the brain has capacity to regain function through the creation of new neurons (neurogenesis) and new neural connections. Similarly damaged chimpanzees cannot recover.

Lifelong youthfulness is not all fun and games, however. There is a trade-off in staying young. In many ways, wolf behavior is better adapted to survival in a difficult environment. Golden retrievers might not last a week in the wild. Dog behavior, on the other hand,

is more suited to coexistence with humans. One set of behaviors is not really "better" than the other. Neoteny opens the door for greater adaptation, but brings with it certain costs. Neoteny tends to be more flexible but vulnerable, while maturity is stronger but more rigid or brittle.

Neoteny is a boon to humans: it has allowed us to come down out of the trees and live anywhere on the planet. We are designed by nature and evolution to continue playing throughout life. Lifelong play is central to our continued well-being, adaptation, and social cohesiveness. Neoteny has fostered civilizations, the arts, and music. While neoteny has its drawbacks, it's simply how we are built. The psychiatrist Erik Erikson sums it up beautifully: "It is human to have a long childhood; it is civilized to have an even longer childhood. Long childhood makes a technical and mental virtuoso out of man, but it also leaves a lifelong residue of emotional immaturity in him."

Of all animals, humans are the biggest players of all. We have stretched the juvenile development program to a minimum fifteen years. Brain scans and behavioral analyses have demonstrated that in modern society, the executive centers of the brain continue to undergo changes into the twenties, a fact that our alcohol laws respect but driver's licenses do not. But our brains don't stop evolving after our twenties. In an individual who is well-adjusted and safe, play very likely continues to prompt continued neurogenesis throughout our long lives. For example, studies of early dementia suggest that physical play forestalls mental decline by stimulating neurogenesis. Research on this subject is really in its early stages, but there are also a few studies that show a relationship between continued use of puzzles, playful exercise, games, and other forms of play and resistance to neurodegenerative disease.

play in adulthood

How do adults play? The answer is not as obvious as it might seem. Many of the things we regard as play may, on closer inspection, have the qualities of work. And what to many people might seem like work may really be built on a foundation of play. A golf game might be the epitome of play, or it might be part of a calculated, controlled effort to close a big sale.

I live near Pebble Beach and have played on its famed golf course a few times. For most golfers, playing at Pebble is one of the highlights of their life, a special golf moment that they may have dreamed of for years. Yet I've seen golfers who are ticked off when they tee off and feel no different after eighteen holes. They are so miserable and angry that they spread their unhappiness to everyone around. These people are not playing. They are self-critical, competitive, perfectionistic, and preoccupied with the last double bogey. These emotions don't allow them to feel the playful, out-of-time, in-the-zone, doing-it-for-its-own-sake sensation that accompanies joyful playfulness.

Runner's World magazine once divided runners into four types: the exerciser, the competitor, the enthusiast, and the socializer. The exerciser is someone who runs primarily to lose weight, to stay in shape, to improve cardiovascular fitness. The competitor runs to improve race time, to beat others, to make a PB (personal best).

Enthusiasts run to experience the joy of the day, to feel their muscles working and the air on their face. For the socializer, running is primarily an activity to bring people together for talking, which is the real fun.

All four types are certainly running, but the internal experience can be very different. The truth is that the enthusiast and the socializer are most likely to be engaged in pure play—pursuing the activity for the joy it brings (and you could say that for the socializer the source of joy is the talking, not the running itself). The other two may be running mostly in pursuit of goals—perhaps fast times or fitness—that can take away the joy from the experience and add stress to their lives. If exercisers or competitors feel lousy when they don't meet certain expectations they have for themselves, what they are doing is not really play. On the other hand, the thrill of competition may be a necessary and healthful part of the competitor's play.

Sometimes running is play, and sometimes it is not. What is the difference between the two? It really depends on the emotions experienced by the runner. Play is a *state of mind*, rather than an activity. Remember the definition of play: an absorbing, apparently purposeless activity that provides enjoyment and a suspension of self-consciousness and sense of time. It is also self-motivating and makes you want to do it again. We have to put ourselves in the proper emotional state in order to play (although an activity can also induce the emotional state of play).

Neuroscientist Jaak Panksepp, who extensively studied play in rats and other animals at Bowling Green University and now has a play research center at Washington State University, believes that play arises first in the human brain stem, where survival mechanisms such as respiration, consciousness, sleep, and dreams originate. This initial activation (which is in-built and hardwired) then connects to and activates pleasurable emotions that accompany the process of playing. Without this emotional linkage, what occurs is something other than play.

Watching sports, sitcoms, *Oprah*, or an excellent drama on TV is usually a type of play, as is reading a novel. Think about how you feel walking out of a really good movie, bringing your mind back again to the everyday world but retaining a changed perspective. One critic remembers walking out of *Lawrence of Arabia* and feeling that the sunlight looked different. This sense of coming back to the world shows that the movie was indeed play. So is reliving its scenes in your mind later. Hobbies like model airplane building, kite flying, or sewing are most often play.

In fact, I would say that the impulse to create art is a result of the play impulse. Art and culture have long been seen as a sort of by-product of human biology, something that just happens as we use our big, complex brains. But the newer thinking is that art and culture are something that the brain actively creates because it benefits us, something that arises out of the primitive and childlike drive to play.

If we look at a life over time, and observe the origins of many artistic expressions, they are rooted in early play behavior that gets encouraged by natural talent and richness of opportunity in the environment. Watch a two-year-old who is drawn to music spontaneously dance to the beat of a summer band concert in the park. Fifteen years later, that kid may be a consummate pianist or just spend hours humming and strumming a guitar. But the draw to rhythm and music were kindled by spontaneous playfulness when the band started playing during that long-ago summer. The emotions that fostered this embrace of music were not verbal nor a product of thoughts like "I think I'd like to be a musician." They were prompted by a deeper, more primal process, which I believe Jaak has captured in his descriptions of processes that link brain stem (movement) to limbic (emotional) to cortex (thought).

The same could be said for what draws anyone to painting, athletics, poetic rhyming, punning, being a mimic, organizing the furniture, loving paper airplanes, and so forth. Each of these play-based potentially aesthetic outflows arise from preverbal, emotionally inspired variety offered to an open heart ready to play.

Another line of thought is that art promotes community integration and interaction. Music, dance, and painting, so often part of harvest festivals and religious observances, bring people together to "sing with one voice." Art is part of a deep, preverbal communication that binds people together. It is literally a communion.

This "belonging" is an outgrowth of early social play among kids. Getting in sync with local groups of kids, and being able to follow that lead into more complex communal groups is a necessary ingredient for cohesive community life when conflicts and differences of style and opinion must be hammered out. This applies to families, school boards, church committees, and, unfortunately, to Congress, which appears to have a major play deficiency in this area of socialization.

For adults, too, taking part in this play is a way to put us in sync with those around us. It is a way to tap into common emotions and thoughts and share them with others. The Australian director Baz Luhrmann expressed this well when he was instructing drama students in how to perform Shakespeare. His students were speaking their lines flatly, trying too hard to get the words right, and hearing too little of the emotion in the script. "In the end it's called a play because [we] play," Luhrmann told them. "It's a game. It's a big, fun, silly, but profoundly moving, human game."

We don't participate in community dances much anymore, and our entertainments are often created by multimillionaires who live far away, but the effects of play continue to filter through society and foster community. We follow the thrill of victory and the agony of defeat on ESPN. We experience Mafia life with *The Sopranos* and talk about it with our friends.

What about work, which is supposed to be the opposite of play? Is there play in something as serious as biochemical research? After taking play histories of Nobel laureate scientist Roger Guillemin and polio researcher Jonas Salk, I realized that what they were doing every day in the laboratory was playing. When Roger took me through his laboratory he was like a kid as he described his experiments. Here was the biggest, most expensive sandbox he had ever played with, all set up to let him discover wonderful new things. I still remember his glee when he told me about his latest work: "Releasing factors, Stuart, we have discovered releasing factors." His joy was as pure as that of a kid showing off a beautiful shell picked up at the seashore.

Their play was esoteric and difficult to understand for most people, but activity and the joy were the same as they would be for a kid in a sandbox. For Salk and other scientists I talked to, the thrill of their play was amplified because their particular sandboxes were filled with expensive toys, and the implications of their research were so profound. It wasn't just Mom applauding about a well-built sand castle, it was about making a difference in the world, as well as being on the cover of *Time* magazine and receiving a medal from the king of Sweden.

"Work" that is actually play can be found at a much less lofty level, too. I once met someone who said the best job he ever had was working in his family's auto junkyard. He said that he and his brother would compete to see who could sort through a stack of used carburetors fastest, or break down and rebuild generators. They got a lot of work done, but it was also great fun.

The work that we find most fulfilling is almost always a re-creation and extension of youthful play. The engineers that JPL found to be so adept were the ones who had played with their hands in their youth, taking apart clocks and cameras, building forts and stereos. They performed well as adult engineers not because they had lots of practice working on watches, but because in a sense they were doing for work what they had always done for pure enjoyment. They were still playing.

A recent cover of the *New Yorker* magazine captured this playful aspect of engineering for me. It showed a guy high above New York on a skyscraper under construction, sitting on one bare I-beam and drawing the next one, creating the building by drawing it. It brought to mind the famous children's book *Harold and the Purple Crayon*, about a boy who creates his own world and gets in and out of scrapes by drawing purple lines on the blank book pages. The guy on the *New Yorker* cover even looked like a grown-up version of Harold, now inventing his own world on an adult scale. I think that for architects and engineers who are really good at what they do and enjoy their work, there is always a strong echo of the enjoyment they got as they designed and built things in their childhood.

what is your play personality?

As we grow older, we start to have strong preferences for certain types of play over others. Some things float your boat, others don't. Over the years, I've observed that people have a dominant mode of play that falls into one of eight types. I call these play personalities. These categories are not scientifically based, but I've found them to be generally accurate.

No one is a perfect example of a single play personality type; most of us are a mix of these categories. At different times and in different situations, people might find themselves playing in a mode that is different than their dominant type. I've found that most people recognize themselves in these archetypes and find them useful for discovering their own play personality. This chapter will discuss the types of play personalities and then offer readers a chance to identify their own type as a way to achieve greater self-awareness and greater play in life.

The eight personality types are:

The Joker

The most basic and extreme player throughout history is the joker. A joker's play always revolves around some kind of nonsense. Indeed, nonsense is the first type of human play we engage in: all baby talk begins with nonsense. Parents make infants laugh by making silly sounds, blowing raspberries, and generally being foolish. Later, the class clown finds social acceptance by making other people laugh. Adult jokers carry on that social strategy, though. George Clooney is a notorious practical joker. On the set of *Ocean's Twelve* and *Ocean's Thirteen*, he and fellow actor Matt Damon reportedly would try to outdo each other with practical jokes.

My dentist, John Lauer, also qualifies for this designation. As the dental chair reclines, his patients see a large ceiling sign that says, "The dentist tells the jokes," and inevitably he has a new one on the tip of his tongue.

The Kinesthete

Kinesthetes are people who like to move, who—in the words of Sir Ken Robinson—"need to move in order to think." This category includes athletes, but also others like Gillian Lynne, who find themselves happiest moving as part of dance, swimming, or walking. Kinesthetes naturally want to push their bodies and feel the result. They may be those who do football, yoga, dance, or jump rope. While kinesthetes may play games, competition is not the main focus—it is only a forum for engaging in their favorite activity.

A friend of mine has a daughter who is a gifted young gymnast. Her academic parents became concerned over the difficulties she had in reading comprehension. Her mother, however, noticed that she could concentrate on tasks when she was bouncing on a large inflated ball. So her mother began to have reading lessons with her as she sat, moving on the ball. The daughter's reading ability and skill at retaining information has been improving consistently. She will likely never choose to become a librarian, but her sense of failure at reading has vanished.

The Explorer

Each of us started our lives by exploring the world around us. Some people never lose their enthusiasm for it. Exploration becomes their preferred avenue into the alternative universe of play—their way of remaining creative and provoking the imagination. Think Richard Branson or Jane Goodall.

Exploring can be physical—literally, going to new places. Alternatively, it can be emotional—searching for a new feeling or deepening of the familiar, through music, movement, flirtation. It can be mental: researching a new subject or discovering new experiences and points of view while remaining in your armchair.

The Competitor

The competitor is a person who breaks through into the euphoria and creativity of play by enjoying a competitive game with specific rules, and enjoys playing to win. He's the terminator. She's the dominator. The competitor loves fighting to be number 1. If games and keeping score are your thing, this may be your primary play personality. The games can be solitary or social—either a solitary video game or a team game like baseball—and they may be actively participated in or observed as a fan. Competitors make themselves known in social groups, where the fun comes from being the top person in the group, or in business, in which money or perks serve to keep score.

The Director

Directors enjoy planning and executing scenes and events. Though many are unconscious of their motives and style of operating, they love the power, even when they're playing in the B-movie league. They are born organizers. At their best, they are the party givers, the instigators of great excursions to the beach, the dynamic center of the social world. At worst, they are manipulators. All the world's a stage, and the rest of us are only players in the director's game. Good examples in this category are *Barefoot Contessa* chef Ina Garten and Oprah Winfrey.

The Collector

What good is a world of random objects? The thrill of play for the collector is to have and to hold the most, the best, the most interesting collection of objects or experiences. Coins, toy trains, antiques, plastic purses, wine, shoes, ties, video clips of race-car crashes, or pieces of the crashed cars themselves, anything and everything is fair game for the collector. One person I know travels the world to see solar eclipses—which might seem like the action of an explorer, except that he has to see every single one and methodically collects evidence of each eclipse. Collectors may enjoy collecting as a solitary activity, or they may find it the focus of an intense social connection with others who have similar obsessions. Jay Leno is famous for his car collection. Collecting cars and working on them is what he does in his free time to play.

The Artist/Creator

For the artist/creator, joy is found in making things. Painting, printmaking, woodworking, pottery, and sculpture are well-known activities of artist/creators, but furniture making, knitting, sewing, and gardening are also in their purview. Artist/creators may end up showing their creations to the world and even selling them for millions, or may never show anyone what they make. The point is to make something—to make something beautiful, something functional, something goofy. Or just to make something work—the artist/creator may be someone who enjoys taking apart a pump, replacing broken parts, cleaning it, and putting back together a shiny, perfectly working mechanism, in effect making it anew. It may mean decorating a room or a house. British prime minister Margaret Thatcher enjoyed wallpapering in her free time. Matisse experimented with many different forms of fine art in his career, from oil paints and pastels to paper cut-outs and stained-glass windows.

The Storyteller

For the storyteller, the imagination is the key to the kingdom of play. Storytellers are, of course, novelists, playwrights, cartoonists, and screenwriters, but they are also those whose greatest joy is reading those novels and watching those movies, people who make themselves part of the story, who experience the thoughts and emotions of characters in the story. Performers of all sorts are storytellers, creating an imaginative world through dance, acting, magic tricks, or lectures.

Because the realm of the storyteller is in the imagination, they can bring play to almost any activity. They may be playing a recreational game of tennis, but in their mind, each point is part of an exciting drama: "The pressure is really on Roger Federer now, he really has to make this point to save the set." In contrast to the competitor, the storyteller's main point of the game is to have an exciting match. Even cooking macaroni and cheese can be transformed through imagination into a worldwide telecast celebrity cook-off. Garrison Keillor and Bob Costas are two examples of natural-born storytellers.

the fate of the sea squirt

If we let the play drive express itself well into adulthood, as we are built to do, we find opportunities to play everywhere. The brain keeps developing, adapting, learning about the world, and finding new ways to enjoy it. Many studies have demonstrated that people who continue to play games, who continue to explore and learn throughout life, are not only much less prone to dementia and other neurological problems, but are also less likely to get heart disease and other afflictions that seem like they have nothing to do with the brain. For instance, various studies have shown that only a small part of the risk of getting Alzheimer's disease is determined by genes. The majority of the risk of Alzheimer's is attributed to lifestyle and environmental influences. One prospective study done at Albert Einstein and Syracuse universities showed that for people who had the most cognitive activity (doing puzzles, reading, engaging in mentally challenging work), the chances of getting Alzheimer's disease were 63 percent lower than that of the general population.

If we stop playing, we share the fate of all animals that grow out of play. Our behavior becomes fixed. We are not interested in new and different things. We find fewer opportunities to take pleasure in the world around us.

From time to time I run into people I grew up with, and like many people I take note of how the years have treated them. As the decades go by, the differences in vitality really show. Everyone notices when a friend's hair turns gray in their forties, and wrinkles really start to accumulate in the fifties, but the friend usually has the same personality and is just as sharp as ever mentally. But I've noticed that the brain really begins to change in the sixties and seventies, and some people start to lose the intellectual sharpness they had before. The people who stay sharp and interesting are the ones who continue to play and work. I have many examples from play histories and my own experience, and I bet you also know a few older people who are interesting because they are still funny and playful.

For example, my friend Allan spent his professionally focused years as a successful pediatric oncologist. He enjoyed a consistent sense that his creativity and compassion offered him a wonderfully rewarding but demanding life. As retirement loomed, he knew he felt vital when he worked with his hands, but he doubted his talents. He began to sculpt wood from large samples scrounged from northern California beaches after winter storms. He is now well into his eighties, has a remarkable sculpture garden and also a wonderful vegetable garden, and spends his days in play. Most of his sculptures depict the indomitable spirit of those kids he treated who inspired him while undergoing chemotherapy or radiation therapy. He tells me he feels sharper and energized by each new sculpture, and his encyclopedic knowledge of gardening gleaned since he was sixty-five feeds his spirit and our need for exotic veggies.

A study done in Okinawa, Japan, by the National Geographic Society revealed that engaging in activities like playing with young children was as important as diet and exercise in fostering the Okinawans' legendary longevity. I remember one particular guy from my own trip there, a wood-carver who was reputed to be close to one hundred. Through a translator, he told me that he laughed all the time as he was carving. I bought a small wooden Buddha from him and asked him what was the longest time he spent on one carving. Two years, he said, and laughed.

When we stop playing, we stop developing, and when that happens, the laws of entropy take over—things fall apart. Ultimately, we share the fate of the sea squirt and become vegetative, staying in one spot, not fully interacting with the world, more plant than animal. When we stop playing, we start dying.

Part Two

living the playful life

Chapter Four

parenthood is child's play

Little Leo is a cherub. At eighteen months, he is bursting with glee and laughter. He clearly and exuberantly loves who he is. His true feelings are written in his every move, unmistakable in his body actions, voice, and expressions. He is fully "here," and the contagion of his joy sweeps his parents and those he meets into a shared state of rapture. Watching him, I feel that "this is as good as it gets." Leo is natural, he is real. While playing together, he and I form a unit that feels fully alive.

Watching Leo at a local child-care center and hearing the peals of laughter from the caretakers as they encourage his playful idiosyncrasies, I wonder, What is he up to? At this age, he loves to push chairs around, move them all over the room. He rearranges those already in place at a children's table, runs them around the room, squealing as he goes, and with his usual intense and happy engrossment. He continues this for a few minutes, then beams and takes a seat at the table, after which he is "through." Like most kids his age, he thrives on repetition and ritual, but this is his very own playful quirk. It demonstrates, I believe, a core in-built play nature, which will, if left unfettered, gradually morph into his "play personality." (The happy director?)

Where does this exuberance come from? And where does it go? At some point in life, many of us lose it. We grow out of childhood and leave behind "childish things." We feel that we shouldn't act this way anymore, and get a sort of willful amnesia for pure play experiences. Seeing a child like Leo reminds us of the joy we may be leaving behind. I talk a lot about play in children, but it's interesting to me that the focus of the conversation often comes back to play in their parents.

Recently, just after I finished a talk at the New York Public Library, a woman came up to speak to me. She smiled, but behind the smile there seemed to be some tightly wound emotions. She said that I had convinced her that play is important, and said she worried about her kids, ten and twelve years old, getting enough time to play but still studying and working enough that they would be successful in life. We spoke about the nature of success, and she realized that what she was really talking about was teaching them how to become responsible adults who have a playful approach to life, who enjoy life, and have work that excites them.

It soon became clear to me that this was something that was missing in her own life. Like many parents, she wanted to give her kids something she didn't have. This woman was enormously successful, a partner at a major New York law firm. She had done all the right things and gone to all the right schools, but now she sheepishly admitted that she was not all that happy, as if this were a shameful secret. And now she was getting her kids into all the right schools and the right activities, but she worried she was taking them down the path that she had tread.

Conversations about play and kids also go in the other direction. When I do workshops for corporations, the emphasis is on improving employee innovation and collaboration by understanding play, but people often come back and report that the very first effect of my presentation has been to improve how they interact with their kids.

In the end, I get the sense that adults feel that they themselves didn't get it right, that they once had something special and let it go. They don't know where it went or how to get it back, but they would like to give their children more options than they had.

In some ways, the threat to play is even greater than it was a generation or more ago. Parents who had the experience, as I did, of exploring fields or woods in freedom, worry that kids today are spending too much of their time on video games or in "safe" activities. Some people have noted the brisk sales of *The Dangerous Book for Boys*, which promotes the kind of rough-and-tumble, exciting, and slightly risky games and activities that used to be so common before video games and the parentally organized, high-expectation trip to the soccer field at age six became the norm. Schools have evolved into assembly lines for high test scores, where skills are drilled, supposedly all the better to prepare kids for college.

In many ways, these developments give kids a more privileged, more sophisticated view of life, and yet we have the feeling that something is lost—perhaps unfettered imagination and freedom.

This chapter is nominally about play in children, but really is about play in all of us. Childhood is where it all begins, the time of life when play comes naturally. It is the foundation of what we do for the rest of our lives. As we understand the unique play patterns of childhood, we can start to see our child' s own truths and purpose, and we can begin to visualize a path that will lead to happiness and fulfillment. Yet helping them along that path requires that we remember the emotional states of play in our own childhood, which provide clues to the primal state of playful bliss for each of us.

On commercial airplanes, the instructions for emergency procedures tell adults that in the event of cabin depressurization, they should put on their own oxygen mask before they assist children. Likewise, in order to help our children we have to recover memories of how we once played, by retracing our own early play footprints. When we do that and create a playful household, everything from education to chores will go better.

Here are the footprints of play, the baby steps we take from the most basic to complex play behavior, and what they do for us.

play at the beginning of life

Play actually begins to have an effect on a child even before birth, at least indirectly. Unlike most animals, humans spend three-quarters of a year in the womb, emerging at birth as a virtually helpless "fetal" being. During gestation, the embryo and developing fetus are subject to strong prenatal influences from the nutrition to the stress levels of the mother. Even in utero, neural circuits are taking shape, circuits that will set brain patterns for the rest of our lives. An expectant mother's play can lower her stress levels and help lessen the discomforts of pregnancy, but that playfulness can also help preform the mind-set of the baby inside her. The effects of the prenatal environment can even be multigenerational. As amazing as it seems, studies of the Dutch "hunger winter" during World War II demonstrate that your IQ, your risk of heart disease, and other health problems are influenced by how well your *grandmother* ate during the third trimester of her pregnancy with your *mother*.

Researchers have also shown that fetal movements—the kicking, punching, or writhing—can also be thought of as an expression of a play program. These seemingly random motions are generated by the central nervous system as a way of making working connections between the limbs and the brain. Once the baby is born, the seemingly random play behaviors will help him explore the world.

Attunement

At three or four months of age, if a child is well fed and safe, and a mother's emotional state is one of openness and calm, when parent and child make eye contact they initiate a harmonic meeting of the minds. As they gaze into each other's eyes, the baby will radiate a compelling smile and the mother will automatically respond with a surge of emotion and verbal and bodily joyfulness—and smile back. He will make little sounds, a babble, or light laughter and she will respond in a rhythmic singsong voice. This is universal across all cultures around the globe.

What's going on in the brain is even more amazing. As they lock

eyes, both mother and child are synchronizing the neural activity in the right cortex of each brain. If we wired Mom and baby and took an electroencephalogram (EEG), you would see their brain currents are actually in sync. This is called "attunement." Their brain rhythms are getting in tune, performing a kind of mind-meld that is a very pure form of intimacy. Fathers, too, experience this as they engage with babies, but traditionally this occurs most between mother and child.

When attunement occurs, both parent and child experience a joyful union. My observation, as I've said, is that this experience is the most basic state of play, and that it becomes a foundation for the much more complex states of play that we engage in throughout life. Allan Schore, a UCLA researcher who has been a pioneer in integrating social, biological, and psychoanalytical theory, has discovered that attunement (also called "bonding") is critical for later emotional self-regulation. Abused children who never adequately experience this end up being extremely emotionally brittle and behave erratically.

The implications of Schore's research and its correlation with other related long-term research on human development are just now receiving the attention they deserve. For example, the child who does not experience attunement because of deprivation or abuse has difficulty forming healthy attachments. This has implications for later adult capacities for stress management and likely affects the efficient working of the emotional regulatory tasks that are generally believed to reside in the right (nondominant) prefrontal cortex. The right and left prefrontal cortices lie in front of the motor centers, and are active in integrating cognitive and executive decision-making. The right side develops more rapidly than the left, is more vulnerable to early deprivational damage, and is felt to be essential to later emotional regulatory activities, risk-taking decisions, and social judgments. If we assume that the neurophysiological models of animal play apply to us, then attunement (the base state of play) buffers the growing infant and child against excessive surges of emotion. It also helps orchestrate the symphony of genetic signals that govern optimal brain development during childhood, adolescence, and young adulthood.

Body and Movement Play

Infants begin to play to make sense of their bodies very early. As I've mentioned, this play program really starts in the womb. Once they are born, the urge to squirm and wave arms continues, and as soon as they can get up on their hands and knees at three to nine months they learn to rock and then crawl. They stick things in their mouths and gnaw with their gums. They roll food around with their tongues, sucking it in and spitting it out, enjoying the process immensely. Later, with spoon in hand, they may catapult or fling a glob of food across the room. These are not random movements—they are intrinsic behaviors that promote exploration and learning. Babbling becomes intelligible words. Babies who are born hearing-impaired will actually use play movements to learn to communicate physically, first babbling with simple symmetric movements, which slowly become simple sign language when an adult signs back consistently.

Movement is primal and accompanies all the elements of play we are examining, even word or image movement in imaginative play. If you don't understand and appreciate human movement, you won't really understand yourself or play. Learning about self-movement creates a structure for an individual's knowledge of the world—it is a way of *knowing*. Through movement play, we *think* in motion. Movement structures our knowledge of the world, space, time, and our relationship to others. We've internalized movement, space, and time so completely that we need to take a step back (a movement metaphor) to realize how much we think in these terms. Our knowledge of the physical world, based in movement, explains why we describe emotions with terms like "close," "distant," "open," "closed." We say we "grasp" ideas, or "wrestle" with them, or "stumble" upon them.

Movement play lights up the brain and fosters learning, innovation, flexibility, adaptability, and resilience. These central aspects of human nature require movement to be fully realized. This is why, when someone is having a hard time getting into a play state, I have them do something that involves movement: because body play is universal. As Bob Fagen says, "Movement fills an empty heart."

The play-driven pleasures associated with exploratory body movements, rhythmic early speech (moving vocal cords), locomotor and rotational activity are done for their own sake; they are pleasurable and intrinsically playful. Yet they also help sculpt the brain. One of John Byers's study of antelopes and other animals revealed that the periods of greatest play were also the time of most rapid growth of the area of the brain known as the cerebellum. If the antelope's activity is inhibited during this period, the growth of nerve cells in the cerebellum is greatly reduced. It has been theorized that the fueling of many areas of the developing brain is enhanced by cerebellar stimulation. While this is a big jump, not yet fully validated by research, it is not illogical. As mentioned earlier, cerebellar functions are now being reappraised by the neuroscience research community. The accumulating evidence confirms that the cerebellum is continuously making contributions (particularly during periods of maximum development) to motor dexterity and to mental dexterity in humans, both of which are required for the emergence of fluent human language.

Object Play

Curiosity about and manipulation of "objects" is a pervasive, innately fun pattern of play, and represents its own "state" (intrinsic pattern) of playfulness. Early on, spoons, teething rings, or foods become objects of play. After a toddler is age fifteen months or so, his or her toys take on highly personalized characteristics. As skills in manipulating objects (e.g., banging on pans, skipping rocks) develop, the richer the circuits in the brain become. We find pleasure in the physical part of object play, in putting together a puzzle, kicking a ball through the goal, or simply tossing a paper wad in the wastebasket. And as lab managers at JPL discovered, object play with the hands creates a brain that is better suited for understanding and solving problems of all sorts.

Imaginative Play

Imagination is perhaps the most powerful human ability. It allows us to create simulated realities that we can explore without giving up access to the real world. The earliest evidence of imaginative play comes at about the age of two in the form of fragmentary stories. Play scholar Brian Sutton-Smith describes these early attempts as partial narratives based on silliness or nonsense without all the elements of a story—beginning, middle, and end. Then, with development, kids acquire the capacity to create a coherent narrative. The imperative to create narrative occurs worldwide in children and is an integral aspect of their play. But whatever the age level or degree of completeness of the story, there is a gleeful verbal experience as the storymaker spins out the story line.

After this, children engage in imaginative play often, naturally and energetically moving freely back and forth between reality and pretend. Determining what is pretend and what is real is usually more important to the adults listening or watching than to the child engaged in the make-believe adventures.

As kids grow older, the line between what is pretend and what is real becomes more solid, but imaginative play continues to nourish the spirit. As I've discussed already, a close examination of adult stream of consciousness demonstrates that the pretend-real process is a lifelong aspect of human thought. We continually make up story lines in our heads to keep the past, present, and future in context. Since kids are embarking hourly on a new life adventure, they use their imaginative urges to keep a context for the emotional and cognitive symphony that is their developing being.

Throughout life, imagination remains a key to emotional resilience and creativity. Deprivation studies demonstrate that fantasizing—imagining the inner life of others and comparing it to one's own—is one of the keys to developing empathy, understanding, and trust of others, as well as personal coping skills.

Social Play

From the simplest game of peekaboo to a formal dance ball, social play has a major role in human play behavior. Humans are social animals, and play is the gas that drives the engine of social compe-

tence. Play allows society to function and individual relationships among many to flourish. Here are a few identifiable subtypes of social play: friendship and belonging, rough-and-tumble play, and celebratory and ritual play.

FRIENDSHIP AND BELONGING

As I noted before, kids begin social play with other kids first through "parallel" play. Two kids may sit next to each other, both playing with sand, water, crayons, or blocks and cognizant of each other's presence, but not interacting directly or emotionally with each other. This type of play serves as a bridge to more cooperative play. Kids playing in parallel are in position to start reaching out to their play neighbor and become part of their game.

By the time they are four to six years old, mutual play becomes the crucible in which empathy for others is refined. As they suggest their own imaginative elements in their game, kids hear other contributions and come to understand other points of view. This mutual play is the basic state of friendship that operates throughout our lives. Give-and-take, with shared contagious enthusiasm, characterizes healthy mutual play.

ROUGH-AND-TUMBLE PLAY

Research on rough-and-tumble play in animals and humans has shown that it is necessary for the development and maintenance of social awareness, cooperation, fairness, and altruism. Its nature and importance are generally unappreciated, particularly by preschool teachers or anxious parents, who often see normal rough-and-tumble play behavior such as hitting, diving, and wrestling (all done with a smile, between friends who stay friends) not as a state of play, but a state of anarchy that must be controlled. Lack of experience with rough-and-tumble play hampers the normal give-and-take necessary for social mastery, and has been linked to poor control of violent impulses in later life. While studying the young murderers in Texas many years ago, we found an absence of rough-and-tumble play in their early backgrounds when compared to similar nonmurderers we interviewed as controls in our study. Since then, the animal experiments with rat rough-and-tumble play, Anthony Pelligrini's extensive studies on playground play, Joe Frost's thirty-eight years of observations on playground play, and my own examination of this form of play through many clinical reviews have all supported the importance of this sort of play. Pelligrini has shown that rough-and-tumble play varies with age. His studies demonstrate that early activities such as chases may relate to later social problem-solving, while the aggressive exploitation that commonly occurs later helps resolve dominance and competitiveness issues.

Joe Frost is a professor emeritus in education at the University of Texas. His observations are more across-the-board, and have resulted in his active participation in playground design features that foster play from early preschool through age twelve. For more than thirty-eight years at the Redeemer School in Austin, he has been innovative in organizing a large playground with multiple free-play environments. From watching and creating these playground opportunities, he sees that there are many lifelong benefits derived in playground settings. He offers young children environments that foster graduated exploratory play,with areas that incite vigorous free play and spontaneous games, so that lightly supervised rough-and-tumble play and spontaneous games are fully experienced. He has also established a verdant playground adjacent to the more traditional one, an area with trees and native plants where solitary play occurs.

While his observations are impressionistic, they are based on systematic long-term observations with informal follow-up. Frost has strong opinions about the value of rough-and-tumble play to provide a necessary and important base for successful cooperative socialization.

In Frost's mind, rough-and-tumble play is generally defined as friendly or play-fighting and may be extended more broadly to any active play that includes body contact among children. In practice the meaning is also extended to superhero play, typically influenced by television characters. Children in school assume roles of "good guys" and "bad guys" and play out themes in such games as tag, chase, king of the mountain, and mock karate. From its beginnings in infancy, rough-and-tumble play is integrative in nature, initially including preliminary symbolic play and organized games, over time expanding into more sophisticated versions, and with experience and development taking on characteristics of organized sports.

"Many adults, including teachers, do not distinguish between play fighting and real aggression, and prohibit any form of wrestling, shouting, and make-believe aggression," Frost observes. "However, children know the difference between friendly and real aggression, and when allowed, engage in rough-and-tumble play very actively, changing the nature of the game to accommodate interest and/or demands by a self-appointed leader."

Outright physical rough-and-tumble play diminishes with age, but a lifelong involvement in games, sports, and group activities that not only tolerate but foster creative tension is a natural extension of this type of play.

CELEBRATORY AND RITUAL PLAY

Children don't need formal events to initiate play, because they do so naturally, but it is important to mention celebratory and ritual play as one of the types of social play. Celebratory and ritual play may be a birthday celebration, a dance, a holiday dinner, or a seventh-inning round of "Take Me Out to the Ball Game." Children don't spontaneously initiate these events, but such ritual social experiences create a reservoir of good memories and help them develop a taste for ritual play as adults.

Serious adult rituals often are accompanied by celebratory play, like the reception after a wedding. In adults, ritual and celebration is often necessary both to provide an "official" excuse to play and to keep this play pattern under social control.

Storytelling and Narrative Play

"The wind howls outside and rain pelts the roof. The little girl's mother said she would be gone for only a few minutes, but now hours have gone by and the night is black as a dungeon. The lights flicker then snuff out. And she hears a horrible moaning from the basements . . ." Storytelling has been identified as *the* unit of human understanding. It occupies a central place in early development and learning about the world, oneself, and one's place in it. A critical function of the dominant left hemisphere of the brain is to continually make up stories about why things are the way they are, which becomes our understanding of the world. Stories are a way of putting disparate pieces of information into a unified context. As we grow, the drama of stories enliven us and the narrative structure tells us something about how things are and how things should be, whether we are listening to Big Bird's take on life or Garrison Keillor's tales of Lake Wobegon.

Stories remain central to understanding well after childhood. When people make judgments about right and wrong, even in politics or the jury box, they often do so as a result of a story that they construct about events that have happened. You may have even noticed, as you read this book, that stories about research or play situations are much more engaging and may feel more informative about the science of play than some of the straight science. It's just human nature.

Storytelling has the capacity to produce a sense of timelessness, pleasure, and an altered state of vicarious involvement that identifies narrative and storytelling with states of play.

Transformative-Integrative and Creative Play

Play can become a doorway to a new self, one much more in tune with the world. Because play is all about trying on new behaviors and thoughts, it frees us from established patterns. For children, who are always in the process of changing and becoming, transformative play is a constant part of their world, and often goes unnoticed.

Sometimes, though, in kids who are really stuck, play can provide a dramatic and obvious example of transformation. In the process of completing the PBS film series *The Promise of Play*, our crew was allowed to follow the progress of a nine-year-old girl who had been identified as depressed, friendless, and playless. At her school, time was set aside for her to work with a group called Positive Intervention Through Play, which was orchestrated by a teacher trained as a play therapist. We observed and filmed the girl over a four-month period. As she began to tap into basic play modes, dance, play with toys in a dollhouse setting, and so on, her isolative and inappropriate social behavior began to change, and her mood lightened. At the last filming, the crew was in tears as she exuberantly and freely played on the playground with friends (who had previously shunned her for her bizarre nonplayful behavior) and was affectionate with her teachers. Her actual engagement in play, not the coaching of her teacher, had transformed her. Through play, her internal narratives had changed from sterile and sad to rich and imaginative.

When we engage in fantasy play at any age, we bend the reality of our ordinary lives, and in the process germinate new ideas and ways of being. For adults, daydreams may give rise to new ways of doing business. Fantasies may lead to new love. Visualization may lead to a remodeled house or a new invention. Creative play takes our minds to places we have never been, pioneering new paths that the real world can follow. Like when Einstein came up with his theory of relativity after imagining himself riding on a streetcar traveling at the speed of light. Or like when a lighthearted group of designers from the IDEO corporation wildly imagine all the ways they can create an indoor dog-exercising machine. In each case, they are using their playfulness to innovate and create.

In a healthy household, these play stages will emerge naturally, bubbling up from inside the child. Parents who provide a loving, safe atmosphere and model playfulness will allow the play drive to express itself. If these elements are not present, children may miss one or more pieces of the natural modes of play. I had the disturbing but enlightening duty of documenting this firsthand.

the life and death of charles whitman

My first real scientific study of play came in the form of a gun-laden student and what was, until the massacre at Virginia Tech in 2007, the worst episode of mass murder ever on a college campus.

On a hot Texas morning in August 1966, an architectural engineering student named Charles Whitman climbed the campus tower at the University of Texas in Austin and started shooting the people below. His first victim was a pregnant woman. With deadly accurate fire, he killed fifteen and wounded thirty-one before being gunned down by a courageous citizen and an off-duty police officer. Police would find out later that the night before, Whitman had killed his wife and mother.

At the time, I was a newly minted assistant professor of psychiatry at Baylor College of Medicine in Houston. Shortly after the shooting, I received a call from the chairman of the Baylor Department of Psychiatry, who was at a conference in Spain. He was calling me because the governor of Texas had contacted him. Governor Connally himself had been shot and narrowly escaped death during the Kennedy assassination. Now he worried that there were Oswalds and Whitmans lurking on every street corner. Connally had demanded an immediate investigation to discover why Whitman had done this and how we could identify people like Whitman before they struck. Connally authorized the use of all funds and resources necessary, including the governor's plane, to find answers. The department chairman asked me to take charge of the psychiatric component of the investigation, which was arguably the most important part.

Most of the immediate speculations (mine included) were that a ravingly deranged paranoid maniac was responsible, but when the shooter was identified there was shock. Whitman was seemingly a loving husband and son; an ex-Marine, he had been the youngest Eagle Scout in the history of the Boy Scouts. Peering deeper, however, we found Whitman's true nature and how he got that way.

When the small army of blue-ribbon committee members met after their exploratory work was done, the expectation was that each discipline, whether toxicology, neurology, neuropathology, graphology, sociology, psychology, and psychiatry, or law enforcement, would have a particular point of view to propound. Yet we came to a unanimous conclusion. What was *not* surprising to most of us was that overcontrol by Charlie's father and the unending abuse directed at Charlie's mother had been major factors in the development of behavioral problems that ultimately led to what became known in the media as the Texas Tower Massacre. But the incredibly thorough investigation of Charlie's whole life revealed a more surprising factor. After extensive interviews with everyone who had entered

Charlie's life, it became clear that the *lifelong lack of play* had itself been an important factor in his psychopathology.

How did this lack of play have a direct role in his ongoing troubles and his homicidal breakdown? At many junctures in Charlie's life, he could not see outside the box that his father had placed him in. The multiple options found in a free-flowing imagination, which occur spontaneously in a naturally playful, safely nurtured child, were not available. The open exchanges that begin in preschool parallel play, the broadening spectrum of give-and-take offered in pick-up games, and the variety of choices that more intricate play provides were not his to experience.

In Charlie's home, the constant mantle of control and fear didn't allow the emergence of normal patterns of play. Charlie wasn't allowed to play outside with other kids. Instead he was forced to stay inside and do something "useful," like practice piano. When Charlie did get away from his father, when he was out with his mother at the grocery store, his father was still there controlling events by CB radio. When friends of the family did come over, the first thing the elder Whitman would do was put Charlie on display, demanding an impromptu piano recital or showing off some other trick he had taught Charlie.

The result was that nothing Charlie did came from within himself. I interviewed his nursery school teachers, who remarked that even when he was young, Charlie could not really play freely. He would watch what other kids were doing and imitate them, but he wasn't really ever "in" it. The parish priest told me that Whitman didn't get what confession was about because Charlie couldn't talk about what he felt bad about doing—he could talk only about what he thought he was expected to feel bad about.

As Charlie matured, his repertoire of responses to the world was narrow, kept within the boundaries of his father's ambitions for him, and closely monitored. A master of outward conformity, he was inwardly seething for years. He sought out no real mentors to break his father's control, so that by the time of the Texas Tower Massacre, he lived emotionally alone, persevering on a path he could not master, driven not by his own desires or needs, and without alternatives. His final (and only) really autonomous action, narrow in scope but devastating in effect, was an attempt to gain some inner relief by acts of murder-suicide, well described in his diaries.

Charlie is obviously an extreme example of parental overcontrol. In most cases, kids left to their own devices will play naturally, and there is some level of parental supervision necessary to make kids feel safe and protected. Allowing kids to play also doesn't mean that there is no structure to their time. Part of the license to play freely comes from being in an environment that is structured enough to provide a feeling of safety, so that the child is confident that nothing bad is going to happen. Part of the freedom to alter the natural order of things ("Let's say this car can fly"; "Let's throw all the Legos in a big pile on the floor and see what we can make from them") is the knowledge that order will be restored again afterward. Preschool kids may not want to stop playing and clean up, but if every day at ten a.m. they are told in a nonjudgmental, nonangry way that it's time to clean up, and a game is made of it, they will feel the emotional logic of reasonable boundaries, and within this consistency find comfort.

It is a wonder to watch someone who is really good at balancing children's simultaneous but opposing needs for supervision and freedom, for order and disorder. My friend Mindy Upton is one such wonderful teacher. She runs Blue Sky, a preschool in Boulder, Colorado, where the kids have a lot of structure, but they are free within that structure. Mindy, like the best teachers and caregivers, can make a game of anything. There is joy, evident even in the cleanup. Good teachers like Mindy also provide inspiring rituals that match the seasons or ages of the children.

In many tribes, the role of providing order and freedom is assigned to the grandmothers. They are not as harried as other adults, and so they can take the time to let children express themselves, but they have enough experience to set safe limits. Even in our society, grandparents are often the ones who have the time to really listen to children. Parents are often busy trying to mold a child into what they think he or she ought to be. Perhaps grandparents are the ones who see us for what we really are and help us grow into that. I remember distinctly my own grandfather taking me out driving in the countryside when I was ten, then pulling over and switching places so that I could try driving a little. Sometimes he would awaken me before dawn, take me into a small town for a pancake meal, and then take me out to a field where he showed me how to safely handle a gun. There was a sense of discovery and wonder when he presented me with his .22 semiautomatic, a gun that "kept him alive for a winter when he was poor." Real guns were taboo at that age—but he gave me permission to be something other than my parents' expectations, which, though not overbearing, were nonetheless more constricted than his. Perhaps this grandparental dividend is why we live so long, long enough to become grandparents, while other animals die soon after the reproductive years end.

Once kids enter school, the importance of free play doesn't end. All of the patterns that induce states of play are present and remain important for growth, flexibility, and learning. Unfortunately, we often forget this or choose not to focus on play's necessity under intense pressure to succeed. No Child Left Behind is a perfect example. While it is an admirable (and even necessary) goal to make sure that all children attain a certain minimal level of education, the result has often been a system in which students are provided a rote, skills-and-drills approach to education and "nonessential" subjects like art and music are cut. In many school districts, even recess and physical education have been severely reduced or even eliminated.

The neuroscience of play has shown that this is the wrong approach, especially considering that students today will face work that requires much more initiative and creativity than the rote work this educational approach was designed to prepare them for. In a sense, they are being prepared for twentieth-century work, assembly-line work, in which workers don't have to be creative or smart—they just have to be able to put their assigned bolt in the assigned hole.

In fact, Jaak Panksepp suggests that depriving young animals of play might delay or disrupt brain maturation. In particular, his research shows that play reduces the impulsivity normally seen in rats with damage to their brains' frontal lobes—a type of damage thought to model human attention-deficit/hyperactivity disorder (ADHD) because it affects executive functions such as self-control. Panksepp has also performed research studies on normal rats, comparing the brains of those that have just had a major play session with the brains of those deprived of it. In both settings, he and his student Nikki Gordon have found evidence that play increases gene expression in the frontal lobe for brain-derived neurotrophic factor (BDNF), a protein thought to be involved with brain maturation. Without play, Panksepp suggests, optimal learning, normal social functioning, self-control, and other executive functions may not mature properly.

This research has led him to propose a connection between a lack of rough-and-tumble play and ADHD. In fact, based on their findings that "abundant access to rough-and-tumble play" reduces the inappropriate hyperplayfulness and impulsivity of rats with frontal lobe damage, he and his colleagues propose that a regimen of social, boisterous play might be one way to help children with mild to moderate ADHD control impulsivity (and it also is good for those not necessarily prone to ADHD).

learning and memory

Learning itself is enhanced by play, as many teachers know—which is why classrooms often use role-play or simulation to teach a subject that is difficult or perceived to be boring. Kids think history is dry as a bone if they are forced to memorize dates and names, but let them play a game of Diplomacy or imagine themselves living on the frontier in colonial times and history comes alive. Really good teachers also know when to use humor and irony to get lessons across.

My late uncle Bruce taught middle-school science, social studies, and math in Greeley, Colorado. He would wear a duck call to the first day of class, and when asked what it was, said, "It's a secret and has secret powers." Over the next weeks he would discuss duck ecology, migration patterns, and so on as part of the curriculum, continually wearing the duck call but never using it. Then, on the opening day of duck season, he would station his son outside the class with a bunch of tame but wild-looking mallard ducks. Without warning, my uncle would put the call to his lips, blow as if playing a fanfare (*Aaack! Aaack! Aaack!*) while his son threw the mallards into the classroom through an open window. Thirty years later, the head of the Colorado and Pacific Crest Outward Bound schools, who was a member of this class, still remembers the migration patterns of mallards, their winter habitat, and many other details that students of less gifted teachers probably learned but have long forgotten.

Some may cheapen these methods by saying that these teachers are just entertaining students, but what is wrong with that? As long as the lessons are learned as well or better than they would be with other methods. Play isn't the enemy of learning, it's learning's partner. Play is like fertilizer for brain growth. It's crazy not to use it.

As we grow older, we are taught that learning should be serious, that subjects are complicated. These serious subjects take serious study, we are told, and play only trivializes them. And yet learning all the complications of a subject first can be confusing and dispiriting. You may like music and like the sounds you can get out of a piano but are told that first you have to learn the diatonic scale cold. Sometimes the best way to get the feel of a complicated subject is to just play with it. That's why kids often learn computer systems faster than adults—they aren't afraid to just try stuff out and see what works, whereas adults worry that they will do something wrong. Kids don't fear doing something wrong. If they do, they learn from it and do it differently next time.

Learning and memory also seem to be fixed more strongly and last longer when learned in play. While this can be objectified in animal tests, it is also a reasonable hypothesis for humans based on performance and outcome results in a variety of educational settings. The link between adequate recess time and later higher performance is one finding that appears to support these benefits. This may be because of the total involvement that play often requires. The state of play is one in which attention is focused exclusively on the pleasurable play activity, and memory fixation has been shown to be closely related to heightened attention and emotional rewards. In addition, play involves multiple centers of perception and cognition across the whole brain. As psychologist Stephen Siviy, my colleague and National Institute for Play Science adviser, says, "Play just lights everything up" in the brains of rats at play. Siviy has shown how bouts of play particularly affect the brain's levels of certain "intermediate early genes," c-Fos genes, that foster neuronal excitability and survival. Siviy was surprised by the extent of the activation of these genes in the prefrontal cortices of playing rats. He speculates that by strengthening connections between brain areas that might be weakly connected previously, play enhances the retention of knowledge.

The power of playfulness during learning was grasped long ago by the creators of *Sesame Street*, which has become the longest-running children's show in the United States. In 1969, a group of people interested in creating an educational television program decided to deliver lessons the same way companies sold products: in short, catchy, and memorable segments that were consciously made to resemble commercials. They used humor that appealed to children, but their parodies of popular culture also attracted adults. The producers originally meant to have separate puppet segments and live-action street scenes, but tests showed that audiences were less interested in scenes that only included humans. But they found that if puppets and people were in a scene together, the humans could deliver all sorts of practical messages without losing children's interest. The mixture of fantasy and real, jokes and lessons, has been an effective and winning combination for over four decades now, spawning billions in business, launching many successful feature movies, and attracting guest stars who range from Johnny Cash to UN Secretary-General Kofi Annan.

gifts of play

Here's a scene that parents around the world have witnessed over and over. It's a second or third birthday, and the big present comes out. The excited birthday boy or girl tears into the wrapping to uncover the box, then opens the box to find the perfect gift that the parent has worked so hard to find. Maybe it's the hottest toy on the planet, the one that you have to put yourself on a list to get. Perhaps it's a car or doll that has special significance to the parents, one that they had themselves when they were young. Or it could be an heirloom—the actual tin truck or porcelain doll that Grandpa or Grandma had. Imagine the family chagrin, then, when their little darling is more interested in playing with the box than the toy.

Parents should be happy about such a turn of events. It shows that their child has developed a healthy play drive, one that comes from their own fantasies and desires. The box is a blank slate, something they can transform through imagination into anything they want.

As they grow, kids are often taught out of this imaginative approach to play, at first by parents, who might impart pressure and guilt that they really should be playing with this great toy, or by pervasive media marketing. Later, kids get toys that come straight out of hit movies or TV shows, toys that come with a preset collection of ideas about who the characters are and how children should play with the toys. This kind of preformed script can rob the child of the ability to create his own story. Instead, he is mimicking the expressions and lines that he is expected to say. A chance for imaginative flights of fancy is lost.

Authentic play comes from deep down inside us. It's not formed or motivated solely by others. Real play interacts with and involves the outside world, but it fundamentally expresses the needs and desires of the player. It emerges from the imaginative force within. That's part of the adaptive power of play: with a pinch of pleasure, it integrates our deep physiological, emotional, and cognitive capacities. And quite without knowing it, we grow. We harmonize the influences within us. Where we may have felt pulled in one direction by the heart and another direction by the head, play can allow us to find a balanced course or a third way. All evidence indicates that the greatest rewards of play come when it arises naturally from within.

When play arises out of innate motivations it is also likely timed to occur when we are primed for the most synaptic neural growth.

That is when we are embracing the issues that grab us most, the ones we may not even be able to voice logically. Our desires and needs are preconscious, inchoate, and the act of play gives them form and breathes life into them. This process is easier to see in the young, but occurs throughout life. You can see it in the glee that a young boy gets from watching a fire truck go by, and then see it once again as he voices his urgent need to go to the fire station, get a toy fire truck, or put out imagined fires on the family room floor.

We may think we are helping to prepare our kids for the future when we organize all their time, when we continually ferry them from one adult-organized, adult-regulated activity to another. And, of course, to some degree these activities do promote culturally approved behavior as well as reinforce our roles as "good" parents. But in fact we may be taking from them the time they need to discover for themselves their most vital talents and knowledge. We may be depriving them of access to an inner motivation for an activity that will later blossom into a motive force for life.

It used to be that self-organized play was all kids did. Most adults over the age of forty-five will likely have memories of exploring on their own, through puddles and fields or on city streets. The only direction they got from their parents was to be home for dinner or before dark.

The pick-up games of my own youth were typical of those that spontaneously cropped up in vacant lots and parks across the country. They were anarchic, and didn't always end well, but they had their own style and etiquette, were full of interruptions, reversals, flexibility, and rule changes. Despite the seeming anarchy, these games existed within an overall, agreed-upon sense of structure, and fairness. They were undertaken with an accepted, minimal risk of damage, and had safeguards. Although there was considerable mayhem and noise during the course of these games, I remember that they were exciting, and there wasn't naked aggression.

These games certainly let me know who I was: feisty, vulnerable, a fairly fast and shifty runner, but not as gutsy a tackler or as courageous a blocker as my brother. I needed his protection and that of his friends to take on the challenge of the games, and they freely offered it. It was terribly important to get in the game, and *belong*. I remember feeling . . . fierce. But it was okay to cry if you got hurt. Not okay to cheat or whine. Not okay to make fun of one of your own team members, but okay to ridicule someone on the other team. If team members changed in the middle of the game, it was then okay to try to verbally humiliate the former teammate.

After these games we discussed the big plays, the great plays, the incredibly lucky plays. We had our own verbal highlights reels, letting kids shine for a few moments in the spotlight for their abilities. We also let the screwups know their deficiencies in clear language.

I can attest that these child-organized, child-dominated experiences have had major effects on not only my own capacities and perceptions as an adult, on how I grew to see myself, but did so also for all the other boys who were involved in these games. Dougie Weaver was a star, and became a Big Ten halfback, and later a major coach. The other star, Linnie Keith, faded athletically as he started high school, but his academic interests and tinkering nature led him into dentistry.

Looking back now, these play-inspired portraits of me and others feel more real, persistent, and predictive of life to come than any other molding experience. Certainly, parents and mentors are pivotal, but the self that emerges through play is the core, *authentic* self.

Alas, these kinds of activities are becoming rarer. Most suburban kids now are ferried from music lesson to math tutor to soccer game. Every activity is organized and overseen by adults. This is not all bad. I think that in many ways the relationship between kids and adults, kids and their parents, is much closer than in my day. But something very significant has also been lost.

My last trip to suburban Chicago and the old neighborhood revealed no empty lots, no pick-up games, a lot of adult-supervised youth sports, busy kids and parents, many more and much nicer cars, less street noise, clearer air, mixed ethnic groups in sidewalk restaurants, and multiple families living in the big Victorian houses. Not the same world. Talk to kids on the street (when they can be found—they are usually leaping from a car to their next adventure) and they are generally more hip, glib, are inevitably texting and on cell phones, and are more comfortable with adults than I remember any of my gang of buddies.

On the other hand, I believe that by the age of ten or eleven, kids, even on the soccer and Little League fields today, create their own, private play domains, which may seem like just goofing off when the coach isn't looking. Nature's design for play is just too strong to be pushed aside completely. They will find their own new ways of asserting their own community, socialization patterns, and individuality. One journalist recently wrote about driving his kids and their friends from one activity to another, when he noticed a strange giggling among the group in back. Upon further inquiry, he discovered that they were all texting one another on their cell phones so that they could "speak" frankly, but secretly, in the presence of the adult. They were creating their own private play zone where they could socialize freely.

It's easy to start to worry about risks when kids create their own play. What are the kids texting each other that they don't want me to hear? Is it profanity? Inappropriate sexual talk? Are they hurting someone? Or if I let them freely roam, are they going to ride their bike in the pathway of a car? Will they drown in the pond? Start taking drugs? These are the kinds of questions that parents have wrestled with since antiquity and they will probably continue to do so into eternity. But part of being a parent is learning to accept the limitations of our ability to make our kids safe, successful, and happy. We should strive for all these things, but eventually they will grow and learn to be on their own (we all hope). All parents need to foster that internally driven, self-directed play that will allow children to become secure and self-confident on their own. There are risks to this sort of play, and the risks should be monitored and minimized. But trying to suppress free play or rigidly control kids' activities poses, in my long-term experience, a far greater risk to their future health, success, and happiness.

coming into adulthood: sophomoric rites of adolescence

The Greeks perfectly captured the paradox of adolescence in the term "sophomore," which literally means "wise-foolish." In adolescence, kids are pulled in opposite, incompatible directions. They are torn apart. Adolescents are expected to take on adult responsibilities and yet not given adult privileges. They can be amazingly perceptive, sometimes even more so than the adults in their lives, and yet they can still make the most fantastically colossal errors of judgment.

In the end, the goal of adolescence is not only to acquire the skills to thrive in a world in which we are mutually dependent, but also to acquire a sense of individuality and uniqueness. Kids have the task of separating themselves from their parents while maintaining a close and affectionate relationship. No wonder it is a difficult and confusing time.

In order to do this, adolescents grow a new brain. I exaggerate only a little.

Neuroscientists have shown that during puberty, a whole new set of brain genes that have been silent since birth turn on, creating a flowering of new neural growth and pruning of the cortical neuronal trees at a level unmatched since our early development in the womb. As the neural tangle works itself out, kids can see the world in unique and surprising ways. Studies have demonstrated that adolescents who are shown pictures of various facial expressions will often make very odd (and wrong) inferences about the emotions the people in the picture are feeling. Because of these odd perceptions of everyday stimuli, teens in some ways are living in a different reality from the rest of us. And it doesn't just happen during the teens. This brain growth continues well into the twenties. This is especially significant as our society extends adolescence out beyond the traditional high school years.

The adolescent period often is a time in which kids are pushed to be serious, to "put away childish things," and prepare for the adult world. And yet I would say that learning how to stay playful in age-appropriate ways while taking on those responsibilities is one of the most important tasks of this age.

What does that mean, really? How is it possible to *remain* playful when so much must be done? Kids have homework to complete, chores, sports programs. In many families there are also jobs after school, responsibilities at church or temple functions, or other family obligations. Some kids, when asked if they are overloaded, will look at this list and say, "Check, check, check to all of the above."

The need for kids to work hard and achieve, while at the same time playing and finding their own joy, comes to a head for many families as the kids get close to college age. This is never an easy or simple subject to address. Parents feel responsible for providing the best opportunities possible for their kids. Providing a prime education and helping them to accumulate an impressive résumé by the end of the junior year in high school is the norm for many families. However, this pattern is "normal but not healthy," as my playful friend Dr. Bowen White says. What are the alternatives? I advise allowing wisely guided personal choice, initiated by your child. Provide plenty of environmental opportunities early (starting much earlier than high school) and encourage early play patterns that have risen directly from the natural choices that your child's early play demonstrates. The parents of Gillian Lynne, the dancer and choreographer, spotted an early talent (that at first looked like a big problem) and encouraged it. Parents and educators, corporate leaders, and others need to become convinced by the evidence that long-term life skills and a rewarding sense of fulfillment—and yes, performance—are more the by-product of play-related activities than forced performance.

True mastery over a lifetime comes from one's internal play compass. When parents and teachers push too hard to get kids to perform, children do not experience feelings of competence and do not create from within their own sense of mastery.

Whether the pressure to get into a premier college is prompted by parental ambition, cultural pressures, or economic necessity to remain upwardly mobile, the result is that many high school kids give up everything they enjoy in the pursuit of résumé perfection. Harvard alumnus Michael Winerip writes about interviewing a bright young applicant to Harvard for the school's admissions committee; despite the student's 4.0 GPA, high SAT score, volunteer work in the local hospice, forgoing summer vacation to teach reading at the local prison, etc., etc. (in other words, a perfect résumé), this would be another kid who most likely wouldn't be admitted because there are so many of these "perfect" kids applying. Parents know this, and the expectation is to be even more perfect. The pressure on these kids is extreme. *The New York Times* recently published an article on a high school where kids were going without lunch in order to maximize their academic credits as a means of adding to their credentials for admission. As an article in *Slate* stated recently, "Those who play the game most intensively are often rewarded: The child who takes fifteen AP courses, plays the clarinet in three orchestras, runs a Cambodian refugee camp in the summer, and eschews lunch all winter really does have a better chance of getting into college than the child who plays kickball after school in the empty lot next door."

What this demonstrates is the lack of appreciation for what freely chosen play-based activities contribute to long-term life satisfaction. For a dozen years, I have been presenting a seminar on play every fall to a group of preselected Stanford sophomores, followed by their participation in a two-week leadership immersion. Something I have noticed is how consistently bright these students all are, but as admission to Stanford has become more competitive in recent years, I have also noticed that their sense of autonomy has diminished. They appear to me, at least, to initiate less and less spontaneous joy than in years past. They seem to be in command of more information, combined with a consistent radar homing them in to be *pleasing* to the professors. With a few exceptions, they are, in my opinion, suffering from chronic low-grade play deprivation, and are so used to their hectic, pressured, high-performance lives (despite still being kids) that they don't realize what they have missed in the pursuit of academic excellence and success.

Perhaps a healthier trend is being set by Naperville School District 203 in Illinois. In his groundbreaking book *Spark*, John Ratey writes how Madison Junior High School physical education director Phil Lawler started a revolution in the school's approach to physical education. The model developed in Naperville focuses on aerobic exercise and lifelong fitness, giving kids the skills and experiences necessary to lead physically active lives. The school takes an individual approach, encouraging each student to progress toward his or her personal fitness goals. Students engage in a wide variety of physically strenuous play, involving traditional games and activities, as well as more novel play activities, such as climbing walls, Dance Dance Revolution, and a video game–like interactive stationary bike course simulator.

PE4life, a not-for-profit organization dedicated to promoting quality physical education in U.S. schools, has adopted the Naperville model and is now exporting it by setting up PE4life academies in schools around the country. By allowing "choice" of activities, they are introducing play as an integral dimension of academic programs. Naperville's District 203 high school has been opening up new vistas by continuing the middle-school programs, modified for older kids. Academic performances have increased dramatically as fitness has improved. Phil Lawler started by introducing cardiovascular fitness, even to nonathletes, and was surprised to see grades going up. As the health and academic successes began to accumulate, more playful activities have been introduced, all very active, including square dancing.

So instead of padding their résumés with more academic courses and shunning PE, Naperville has, in my judgment, combined hard physical activity with academics and play. While monitoring cardiac fitness is a major component of these programs, the evidence is overwhelming that by allowing kids to *choose* their activity (it does have to be hard aerobic exercise) the kids are having fun, improving their bodies and also their brain function.

I think it is important for kids to keep a sense of perspective. It is important to recognize that taking care of responsibilities, getting good grades in school, and all the other teen duties are important, but they are not the be-all and end-all of life. These things are all, paradoxically, important but not important. This sense of having a little distance, a sense of irony, is what gets us through the hard times. Adults can model these qualities and improve their own life in the process.

By a sense of irony I don't mean cynicism, which comes out of a darker view of the world. I mean the kind of ironic view that allows kids to laugh at themselves and the (often) ridiculous world they live in. Whenever I see a kid with a strong sense of irony, I think, "That kid is probably going to be okay." A playful attitude about life—not really taking everything like popularity or competitive academics or adult criticism so deeply seriously—is key, while at the same time tending to the necessities of growing up, staying within boundaries of the law, taking no inordinate risks, avoiding addictions, and so on. I would recommend a rarely read but classic book here, Joseph Meeker's *The Comedy of Survival*, as a prime text for orienting adolescents (and adults) to a nonidealized world.

A great exercise that I often used for both kids and adults was to ask them to visualize their lives five or ten years in the future, focusing not on whether they want to be a lawyer or be rich, but instead on what they might be doing that would make them really happy and excited. This itself is a kind of imaginative play. It allows people to step out of their everyday life and see the bigger picture. It also allows them to see clearly how they really want their lives to be, to focus on those elements that arise out of the true, core self rather than on what the world wants them to be. Once people have this sort of vision, they are empowered to move toward making it a reality.

I think that nobody should keep an ironfisted grip on any particular vision. As with any play activity, people need to be open to improvisation and serendipity. Remember that the point of this exercise is not to create a five- or ten-year plan for personal development. The point is to clarify who you are and what sort of future speaks to you emotionally. The most rewarding activities and interests come to light when we open ourselves up to them. Goals are good, but overly rigid pursuit of those goals can inhibit growth and understanding. After discovering what appeals to us emotionally, the strongest emotions, or those that resonate most, lead to the creation of a realistic path forward.

Bill Gates left Harvard because he loved computers. Nate Jones, who showed the importance of hand-play to JPL, liked cars. Oprah recognized her talents and grew with them. Ozzie Smith, the retired baseball star, as a kid threw balls over the roof and chased them down, imagining himself in the majors. These were emotion-laden choices, not cognitively laid-out ten-year plans.

Although I think that sports and other adult-organized activities can be overdone, sports can be a potent training for a playful life during the teen years. Sports provide a ready peer group, united in a common goal. Sports teach how to struggle against adversity, even when the odds seem insurmountable. Adult-organized sports don't have to be antiplay when they are done right.

When I think of model sports programs, I often think of a coach, Gary Avischious, who coaches hockey for six- to nine-year olds. He puts his teams together randomly, he doesn't recruit for talent and as a result gets a real mix of abilities. He starts kids out not with games but on Rollerblades, skating on sidewalks and empty parking lots around town with their hockey sticks over their heads. He has them skate on a hill with a lawn so that they can fall down and not hurt themselves. They have fun and learn to skate around objects and skate backward. This allows them to build their skills, but it also allows Gary to get a feeling for the kids' personalities and spot the ones who need extra help. They also do exercises that bring the kids together as a team.

When they finally do get in the rink, they don't use a puck, but a midget football. It's almost impossible to control, so that they can get the feeling for moving and passing but don't feel shamed if they mess up. When they do use the puck, they are more used to passing and find the puck much easier to control. They are required to pass, pass, pass the puck multiple times before they try a shot on goal during the scrimmage. By this point in the training, the kids have come to realize that all they are ever going to get from their coach is positive feedback. By the time they are finally ready to play actual games, they have learned to work together as a team with energy and skill, and look forward to the fun of competition. The recognition of personal improvement, not stardom, is the gauge for mutual respect.

The result is that Gary's teams have won the championship in their area for thirteen of the past fifteen years. But more important than that, win or lose, the kids love to play. I went to a banquet after the championship game one year when they did not win. Both Gary's team and the championship team were at the banquet sitting at an adjoining table. The contrast between the two teams was remarkable. Gary's team was having a great time, laughing and joking, while the other team was more subdued, serious-looking, its members not looking as if they were enjoying their championship win. The winning coach was pretty hard-core for this age group, always wearing a tie and putting up with no silliness, running the team through skills and drills constantly. They were winners, but anyone looking at the two teams would have thought that Gary's kids were the champions. Indeed, when the waiter came up to Gary's team, he asked them how it felt to win the trophy. They happily told him they had lost.

Athletics provide feedback about one's own physical talents, and what it feels like to participate, win, lose, and be fair. And because sporting contests are games, because the outcome doesn't (or shouldn't) materially affect our well-being, they give us perspective on the other struggles we have in life. No wonder our society is sports-obsessed. William Bowen, a former president of Princeton University, once did a large statistical study to determine if the special preferences that athletes got on college admissions (lower SAT score cutoff, extra financial aid) were unfair. Bowen was surprised to find that, as a group, the athletes actually did better financially after college than other students, a fact he attributed to the drive and energy that sports cultivate. Other studies have demonstrated better mental health later in life among those who have participated in sports (not necessarily varsity level) at college age.

Although as a society we often focus on sports as an essential, group character-building activity of childhood, there are many other arenas in which teens can build a constructive peer group. Speech and debate, drama, math club, art, band, and orchestra also provide a forum for play and exploration among groups of like-minded kids. Home-built-computer clubs used to be big thirty years ago, and now robotics are a big deal among kids in Silicon Valley (the competitions are incredible)—which is just another example of how the toys of today become the technology of tomorrow.

from child to adult

The rites of passage in making the transition from adolescent to adult entail certain changes. Rites of passage involve achieving a sense of self, which brings with it a sense of confidence. They involve breaking away from the norm, making one's own path, facing adversity, and contributing a boon to society. In mythology, the returning hero not only comes back more mature and stronger, but also brings something new that is beneficial to the community. This has been the case throughout human history, in reality as well as in stories and myths.

My own rite of passage involved no dragons but still had plenty of challenges. I got at job at the age of fifteen delivering groceries in a very poor section on the south side of Chicago. My parents knew I was working for the prosperous grocer who lived down our street, but they didn't know any real details of my workdays. On weekends in a mixed-ethnic blue-collar neighborhood dominated by four-story old apartment buildings, I delivered huge boxes of groceries to very large families, mostly on the top floors (no elevators and no heat). The deal that the customers got from the grocer was free delivery if they made a major purchase. In the overcrowded and very smelly apartments lived a real cross section of postwar back-of-the-stockyards Chicago society. There were the dying elderly, the mentally retarded, the alcoholics, the crippled, many newly arrived to Chicago from everywhere. All of them were poor, with the maternal head of household very diligent in having to review every item in the weekly or two-week grocery supply. If I broke an egg, it was a shared tragedy. I spent long days struggling to get to their top-floor tenements.

I hadn't known before this job that people lived like that. That this was often a dangerous part of town, a fact that never fully occurred to me (the witless young knight), but I reveled in having enough strength to carry boxes with bags piled on top up narrow, dark stairways. I dealt hourly with parking hassles, talked police out of tickets, and foiled any attempts to either steal the owner's new Mercury station wagon or my enlarging roll of bills (all deliveries were cash).

This experience opened my eyes. It changed me. I found that I could handle myself in difficult or dangerous circumstances. Of course I look back now and see that I really didn't understand how very dangerous it really was. If my parents would have known, they would not have allowed it. I guess parents of that time did not feel as protective as responsible parents do now. But as I reminisce through the lens of many years of personal experience, I recognize that it was well timed to open me up to actively identify with the struggles and pathos I saw. And though I was just a high school boy delivering groceries in the inner city, this adventure incited me do something more positive with my life than perfect my driving skills. So this rite of passage—exciting, enlarging, and fun overall—continues today in the passion I feel for bringing an enlarged view of play as a boon to others through this book.

For all of us, "entering the forest where there is no path" and discovering our own path is an essential part of the transformative experience.

The Outward Bound program is a good example of a program in which the risks are real but controlled, and the risks always feel much larger to the participants than the leaders. I've seen how such rites of passage can transform kids.

One sixteen-year-old young man, son of my friends, was a major concern to his well-meaning but stymied parents. This kid, Harry, was a very smart and pseudosophisticated couch potato, preoccupied with dark themes in video games, and drawn to ponderous existential literature, but clearly stuck in his juvenility. I had been active on the Outward Bound board, and knew that the program then operative in the mountains of North Carolina was well-led, rigorous, and safe. So on my advice, Harry's parents enrolled him (with only mild resistance on his part). After a hearty breakfast the group of ten city-bred, coed sixteen-year-olds headed off for a distant campsite. They were carrying heavy packs with water, but no food. At about ten a.m. Harry stopped, tired and hungry, and asked for food. The leader said, "Later, much later." For Harry this was the first time in his life he had never been able to snack when hungry. Well, the day was long, deep fatigue set in, but by about four p.m. they arrived at the campsite very tired and hungry. "Here is supper," announced the leader. "Build a fire and cook it!" Whereupon a bag full of live chickens and a sack of potatoes stashed earlier were set in front of the group. Anarchy ensued, as no one had a clue how to kill and prepare live chickens. But hunger and fatigue overrode any reluctance to become primate carnivores. By about seven p.m. the plucked chickens were on spits over the fire, and the famished group ate.

This particular Outward Bound tour, packed with unexpected and challenging (but doable) tasks went on for three and a half weeks, ending with an eight-mile run-walk. Harry made it, completing this experience with a new, confident image of himself. He has never looked back. Now an accomplished triathlete and a medical student, he has been transformed. He credits the mentorship and direct involvement in nature, which began as overwhelming but became fun, for providing the gateway to his successful early adulthood.

Some parents don't allow kids the independence necessary for them to learn self-reliance. I've seen a lot of parents who still write their kids' research papers for them—and these kids are in college. Some of the kids graduate cum laude from top universities, but then are lost. They don't really know who they are because they have never found out for themselves. They are good at interacting with people because they know how to please. Their superiors and former teachers love them. But they feel empty and incomplete inside.

There is one final potential transformation that adolescence brings about, but it is not in the kids. So much of parenthood is just getting by, making sure the meals are balanced and the schoolwork is done, trying to teach responsibility and generosity, right and wrong. But there are times when we pass on knowledge about what really matters in life, about how to look someone in the eye and shake their hands with confidence, about how to have vision, set goals clearly, and have the discipline to attain them. As we adults tell kids these things, we sometimes get a glimpse of our own best selves and how we might live our own lives better.

Part of the joy and pain of being a parent is seeing our own parents in ourselves, seeing their good parts and flaws repeated in our voices. The joys and pains also come from seeing ourselves in our kids, and remembering our own happy days and hurtful traumas reflected in their experiences. If we are not completely full of ourselves or too serious, we can see that we can do a better job of helping our children be more joyful if we help ourselves remember how to play. If we are open to some self-evaluation, and do so with a lightness about our life opportunities, we will actually find a way to play.

Chapter Five

the opposite of play is not work

A colleague of mine, Barbara Brannen, seemed to hit her stride in her thirties. A human resources executive at hospitals and technology firms, she was promoted continuously for years. Happily married, she had one child, and then another. As her children grew, she felt more involved in the community and she began volunteering in a local food bank, became president of the PTA, helped rebuild the Colorado Trail, and became president of her professional association. She spent more and more of her time ferrying her kids to and from playdates or from one organized activity to another. She found herself having less time for the racquetball and skiing that she enjoyed. Just getting the groceries and laundry done on the weekend started to feel like a major accomplishment. She organized get-togethers with friends for drinks and marathon games of Trivial Pursuit, but these too began to feel less satisfying and occurred less often. After a while, she and her friends found themselves so tired that vegetating in front of a movie became the highlight of their downtime. Her social life consisted of talking with friends on the telephone, conversations that were filled with moaning about work, kids, the weather, and life.

It shouldn't have been a big surprise when it all came crashing down. She injured her arm from overwork, and she began to feel a pervasive sadness that she realized came from an inability to live life with the joy she had had before.

JASON STUDIED FINE ARTS in college and became fascinated by the process of making jewelry. He loved the design possibilities, the myriad materials that the jeweler could work with, the fine, detailed craftsmanship necessary to bring a design to fruition. He apprenticed to a master jeweler for a couple years, then opened his own shop in Palo Alto, California. Jason became enormously successful. He would work closely with people to understand exactly what they wanted, design the pieces himself, and then craft them out of raw materials. He was always thrilled with the process of helping people realize their vision, and then being the magician who conjured that dream into reality. But after many years he began to feel a sense of incompleteness, despite the excellence of his work and the acclaim he received for it. He found that more and more, he liked working with people, but the long hours alone in the shop began to feel like drudgery. The more he tried to ignore that feeling, the more the work felt like a burden. Eventually he really didn't want to go into his workshop. He didn't want to look at jewelry at all, which is a severe handicap for a jeweler.

the opposite of play is not work

MARK WORKED in a restaurant right out of high school and loved it. The owner was impressed at how hard Mark worked, but Mark didn't really feel he was working hard. He loved learning about each job in the restaurant and all its systems, and found he had a talent for suggesting changes in the kind of nitty-gritty details that are not glamorous but determine a restaurant's survival. Things like suggesting changes to the menu that lowered the plate price without reducing quality, altering the busboy schedule, and even changing the system for sorting and organizing trash and recyclables, ideas that increased efficiency and profits. There wasn't any part of the restaurant that didn't seem interesting. Every part of the business presented a puzzle he could solve. At the same time, he felt that he was having fun, joking with the kitchen crew and the waitstaff, goofing off in a way that didn't get in the way of the work and made the time fly by.

Soon he was promoted, and then promoted again. As the owner put him in managerial positions with more responsibility, he began to feel that he shouldn't be joking around. He began to feel that it was his job to keep others from goofing off so that they could get more work done. He felt that no one had any right to complain, because he was working harder than anyone else. After closing the restaurant at eleven p.m., he would sit in the office for a few hours going over the bills, then drag himself home with no energy to even shower off the kitchen grease before bed, and then get back to the restaurant before eight a.m. to supervise deliveries. One morning he sat at his breakfast table drinking coffee and watching the second hand tick by on the clock: eight a.m., then nine, then ten. He didn't feel sick, but was so drained that he couldn't imagine mustering the energy to get up from the table. The question that kept running over and over in his mind was so new to him that he thought he was the only one who had ever been stymied by it. But it was one that so many people ask eventually: Is this all there is?

EACH OF THESE PEOPLE, Barbara, Jason, and Mark, is an example of the critical fact that the opposite of play is not work—the opposite of play is depression. Our inherent need for variety and challenge can be buried by an overwhelming sense of responsibility. Over the long haul, when these spice-of-life elements are missing, what is left is a dulled soul.

Far from standing in opposition to each other, play and work are mutually supportive. They are not poles at opposite ends of our world. Work and play are more like the timbers that keep our house from collapsing down on top of us. Though we have been taught that play and work are each the other's enemy, what I have found is that neither one can thrive without the other. We need newness of play, its sense of flow, and being in the moment. We need the sense of discovery and liveliness that it provides. We also need the purpose of work, the economic stability it offers, the sense that we are doing service for others, that we are needed and integrated into our world. And most of us need also to feel *competent*. Even people who are independently wealthy and never need to work a day in their lives find that they need to volunteer or donate to good causes to feel that sense of connection and purpose.

The quality that work and play have in common is creativity. In both we are building our world, creating new relationships, neural connections, objects. Even demolition or sand castle smashing is a kind of creativity, since they clear the landscape, opening the way for new building. At their best, play and work, when integrated, make sense of our world and ourselves.

Respecting our biologically programmed need for play can transform work. It can bring back excitement and newness to the job. Play helps us deal with difficulties, provides a sense of expansiveness, promotes mastery of our craft, and is an essential part of the creative process. Most important, true play that comes from our own inner needs and desires is the only path to finding lasting joy and satisfaction in our work. In the long run, work does not work without play.

It should be obvious by now that play outside of work is undoubtedly a good thing. Play is called recreation because it makes us new again, it re-creates us and our world. As Laurel demonstrated when she began riding horses, just a little true play in one's life can bring everything else, including work, back in balance. A good vacation, one that is not an endless logistical chore of packing, driving, flying, and organizing activities, one that allows us to indulge in activities that we truly love, can also have a transformative effect. When people come back to work from true play vacations, they are eager and energized for work.

Play is nature's greatest tool for creating new neural networks and for reconciling cognitive difficulties. The abilities to make new patterns, find the unusual among the common, and spark curiosity and alert observation are all fostered by being in a state of play. When we play, dilemmas and challenges will naturally filter through the unconscious mind and work themselves out. It is not at all uncommon for people to come back not only reenergized, but also with fresh ideas for work.

All it took for Carl was one day away from the job. Feeling overworked and blocked in his job as an administrator at a hospital, he decided on the spur of the moment one morning to take a vacation day and work out his frustrations on the racetrack. The local Malibu Grand Prix offered unlimited laps in their small, gasoline-powered race cars for twenty-five dollars on Thursdays, and he spent all morning hugging the tight turns and zipping through the straightaways, trying to get his mind completely away from work, watching his lap times drop with each revolution. Around midmorning, though, he seemed to have hit a plateau. His times went up or down a few tenths of a second each lap, but they didn't drop, no matter how perfect he made his line through the turns. So he tried driving really aggressively, punching into each curve, coming out of each turn on the edge of a four-wheel skid, alternatively popping the brakes or hammering down the accelerator as needed. And his times dropped. He didn't objectively feel that the car was going faster, but the emotion that he was driving with seemed to push through invisible barriers.

By early afternoon Carl felt hot and sweaty, but paradoxically cleansed and relaxed. He had successfully rid his conscious mind of any thoughts of his work. But strangely enough, as he pulled into the pits for the last time, he had an epiphany about work. He realized that he had been working too hard to get consensus on every project. "I realized that sometimes people don't want to be part of the decision-making process," he told me. "If someone doesn't have a strong opinion, trying to get buy-in is like stacking Jell-O cubes." Carl realized he needed to punch through the twists and turns of the hospital bureaucracy with the intention and directedness he brought to the racetrack. In solving the problem of his lap times, Carl's playing mind connected with the solution to a problem that was much more important.

play at work

Can play at work itself be useful, though? I would argue that it is *essential.*

First of all, when the going gets tough, the tough go play. Talk to firefighters or police officers and they will tell you that joking around, razzing one another, and practical jokes are just part of the job. Their dark humor often helps them cope with the inherent dangers of their work. If you might die on your shift, it doesn't increase effectiveness to dwell on that fact too much. It's an understandable reaction to become obsessive about the many ways to kick the bucket, but by so doing this preoccupation would most likely cause a freeze-up in the face of danger.

Martha Gellhorn, a journalist who at one point was married to Ernest Hemingway, said that as a war correspondent she got a quick picture of how the war was going in Vietnam when she sat in a mess hall with a group of American officers. Gellhorn had seen combat as a journalist in World War II (she landed in France shortly after D-day) and in the Korean War. The best officers, she observed, had a certain swagger and bonhomie about their deadly work—a joie de vivre or joie de combat. As she sat at the mess hall in Vietnam and mortar rounds began to fall nearby, but not close enough to be dangerous, she saw the officers duck, saw the fear on their faces, and knew the war was lost.

Most work, of course, doesn't hold out the regular possibility of dying on the job. But there are plenty of dangers that feel personally threatening. There is always the danger of looking bad, of being badmouthed, of costing the company money, or being fired. There is the anxiety that comes from the competition, and from market forces beyond one's control. As with many things in life, often the problem is not the problem, the problem is how you react to the problem. If the reaction is that of a deer frozen in the headlights of an oncoming car, the chances of ending up as roadkill are greatly increased. When all employees are focused on the possibility of personal or collective failure, a funereal air saps the energy and optimism necessary for success. At this point, play gives people the emotional distance to rally. When one CEO whom I know gathered employees in the company auditorium to talk about a recent bad quarter, he himself took the blame for the company's performance. He then told the employees that under every seat was a toy dart gun with foam darts, and invited them all to take a shot at him. The air filled with flying yellow projectiles, and the atmosphere of the meeting was completely transformed. The CEO went on to explain how they were going to turn things around, why everything was going to be okay. But the most important point of the talk was nonverbal. The foam darts told everyone that the situation was not so dire. More important, the CEO's playful invitation to take shots at him said that it was okay to have some setbacks and that it was okay to admit failures, take the hit, and figure out ways to fix them. The foam fusillade was a playful prompt to remind the employees of the joie de combat necessary for success.

Sometimes when a situation is really heading south, a moment of imaginative play is the only thing that provides enough distance to see the way out of a predicament. As one high-level executive told me, "Whenever I'm stuck, I try to imagine what someone smarter than me would do, and then I do that."

Andrew Grove tells the story of how he and Gordon Moore used imaginative play to turn around Intel at a critical moment in the company's history. Intel had become a success by making memory chips for computers, yet by the early 1980s the Japanese were fabricating excellent, cheap memory chips, making it difficult for Intel to make any money. There was a need for the company to find another product line to survive, but the company's whole history and identity was tied to memory chips. Nearly all the engineers and sales staff were organized around the production of memory chips, and the company had recently built a massive, expensive fabrication plant to build yet more memory chips. They were stuck in a corporate rut.

One day, Grove and Moore discussed the quandary in Grove's office. They realized that if they didn't solve this problem, the board of directors would fire them and find people who could. They imagined the replacements who would take their jobs and wondered what these superexecutives might do. Grove looked at Moore and said, "Why don't we do it ourselves?" They then "fired" themselves, walking out the office door and walking back in as the better and smarter executives who would replace them. Immediately after acting out that little moment of make-believe, the answer was crystal clear: they had to take the company out of the memory chip business, no matter the cost or internal resistance. Eventually they steered the company into designing and creating microprocessors, for which Intel is now famous.

The paradox is that a little distance from a problem, a sense of perspective, a realization that it really matters little in the end if people choose Huggies over Pampers for their kids, can be one of the most important factors in success. If you are working on a campaign to persuade people to use Huggies, it matters a lot to you and your livelihood that mothers agree with you. If you think too much about your own personal investment in the product, you can become too focused on the real but minute advantages that your product has over the competition (1.5 percent more absorbency, waist fasteners that unstick and stick again ten times instead of eight). An act of imagination can free you up, however, to create other values to the product. The actual war between Pampers and Huggies was won not by judging the qualities of the products, which were similar, but by creating emotional content in advertising that showed a mother with a dry, happy baby. Imaginative play allows people to step back and see both the emotional and the factual elements of a problem.

The beauty of sports is that it embraces the paradox of seriousness and play. We can really care whether the Patriots or the Eagles, the Lakers or the Celtics, win the game, and at the same time realize that it is just a game. If our team loses we may feel down, but there is always a next game, a next season. It matters a lot, but it doesn't matter—which is why sports metaphors abound in business, and why playfulness does not have to involve jokes or goofing off. Sometimes the play may be a friendly competition between teams. Or it can be a very private sort of play-game that the rest of us never see—a personal competition, for instance, to see how fast we can write a memo, or how many things we can check off our to-do list that day. Or it may be the kind of silent narrative that we used to give ourselves as kids at bat: "Now it's the great Mickey Mantle coming to the plate, taking a few leisurely practice swings and settling the bat over his shoulder as he stares down the pitcher." Only now we might channel a personal hero like Warren Buffett, Bill Gates, or your company's CEO. The narrative might not be spoken out loud, or even consciously recognized, but it is still there.

Once we are able to see work as a game full of players, we are better able to accept and make use of a slew of techniques that athletes use to motivate themselves and boost their performance. We might use internal narratives like those I just mentioned, distraction (I'm going to listen to this great music while I work), disassociation (I'm not working, I'm on an island in the South Pacific), or idealization (everyone will be giving me a standing ovation after I deliver this report). All it takes is a little distance. Work matters, but we often allow day-to-day events at work to give us more anxiety than they are worth. Getting oneself into a play state, however, masks the urgent purposefulness and associated anxiety of work, increasing efficiency and productivity.

There's nothing like true play to promote social cohesion at work. When people play, they become attuned to each other. The more powerful players handicap themselves to keep the game going instead of dominating and controlling it. Groups pull together in pursuit of a common goal, which is why the "team" approach is often utilized in business. Team-building exercises often involve play: solving puzzles, building bridges with cardboard, or going through an obstacle course. We talk about *working* together as a team, but it might be more beneficial and productive to talk about people *playing* together as a team. When people graduate from *working* as a team to *playing* as a team, they will really allow themselves to compete fully and with gusto against other teams inside and outside the company.

creativity and innovation

By far the biggest reason that companies want to talk to me about play is its role in creativity and innovation. They want to talk about this because many corporations rightly identify play as their most precious commodity. Production matters now, but creativity is the source of all growth—the new products, new techniques, new services, and new solutions to old problems that mark the difference between a company that will thrive and one that will soon be deader than the eight-track tape.

What's play got to do with it? After all, isn't necessity the mother of invention? The answer to that question is no. I would say that necessity only sets the stage for invention and innovation. Play is the mother of invention. Polaroid may have really *needed* a great new product to replace its lucrative consumer photography business (which is now pretty much dead because of digital cameras), but the necessity of finding a replacement didn't lead to one before the company went bankrupt. (Polaroid's assets were sold to other companies, and the name Polaroid does live on, but much diminished in comparison with its former grandeur.) Necessity is more like a first date. Moving on to become the bride and then the mother of invention requires much more.

What are creativity and innovation? Generally, we think of creation as different than creativity. We can create a snowball or a glob of mud, but people don't generally think of those as examples of creativity. We generally think of creativity and innovation as the production of ideas and products that change our world or culture in some lasting way. They may happen on a large scale, as with the invention of the telephone in the nineteenth century and Internet in the twentieth century. Or they may occur on a small scale, like when someone comes up with an innovative way of organizing files. Creative people create, yes—but what they create shifts the tectonic plates of our world, changing how we think or do things, in ways little or big.

The creative process is popularly thought to be mysterious. Highly creative people come in various temperaments, work habits, and educational backgrounds, making it difficult to find a common denominator to their creative processes. But something so valuable has attracted a lot of thought and study, in the hope that creativity can be produced at will.

Those who study creativity find that the process is by nature contradictory and paradoxical, which is why it can seem so mysterious. Creative people can be simultaneously hardworking and goof-offs. They can have a laser focus on a task, but keep the wide view that lets them see how something fits into the big picture. They are well-versed in their domain of knowledge (painting, physics, literature, the grocery business, etc.), but they don't automatically discount new bits of information that don't seem to fit with the accepted canon. (Nobel laureate physicist Richard Feynman said that when he was presented with a radical new idea about subatomic particles one time, he said, "That's the most ridiculous thing I've ever heard," but before he even finished the sentence he knew that it was probably true.) Creative people can escape into the imagination, but also are firmly grounded in reality. Creative ideas are often those that bring together ideas from different domains or fields.

Many of the paradoxes of creativity are embodied in play. Creative people know the rules of the game, but they are open to improvisation and serendipity. Play activity can be extremely serious, but in the end is "just a game." Much of play takes place in an imaginative world, but is also firmly grounded in reality. In fact, play promotes mixing fantasy and reality. It is designed to activate functionally diverse brain regions to synergistically integrate their function.

Finding people who are innovative and creative is important to a lot of companies. These companies have gathered armies of psychologists and employed batteries of tests that attempt to gauge who is the most creative and why. The recurring problem, as I've said, is that the psychological factors that lead to creativity seem to be all over the map.

As a human resources officer working for a major U.S. bank, Dave Stevens spent a lot of time trying to devise what he called an innovation assessment tool to use on potential hires. He employed well-established personality tests like the Meyers-Briggs Type Indicator and the Minnesota Multiphasic Personality Inventory. Using these metrics, it still took years to find out whether an employee was really going to be a creative force in various departments of the corporation.

Then he heard a seminar I gave through Vincent & Associates Innovation Practitioner Network, and he began looking at creativity through the lens of play. He found accurate indicators for innovation hiding within emotionally charged early play memories. Using a standard psychological test invented for other purposes, he was able to establish criteria to assay whether prospective employees enjoyed novelty, how they reacted to making mistakes and learning from them, whether they were willing to take risks, and other factors that he had not previously identified as being really important. He found that looking at creativity in this way was likely to produce the quality results that he had been seeking for years. He was able to accurately identify creative people and screen out those for whom innovation and creativity were not their primary nature. As a result of these discoveries, Stevens left the bank and began consulting with other HR executives to share what he had found.

Companies have worked to institutionalize creativity in various ways, the mostly widely known method being "brainstorming." As most people know, during brainstorming a group of people is given a problem or puzzle and then encouraged to think up many solutions, the idea being that some of these will be both new and workable. During brainstorming, the group should focus on quantity of ideas rather than quality. In fact, one of the basic rules of the process is that the "quality" of an idea isn't considered at all. All judgment or evaluation of the ideas is put on hold. No one is knocked for what seems to be an off-the-wall or crazy suggestion, and there are no brainstorming mistakes. Sorting through the ideas to figure out the "good" ones will come later.

When the brainstorming process goes well it not only improves the creative process, it also makes people feel smarter, more energetic, more appreciated by their peers. Brainstorming has been credited with more than doubling the productivity of work groups.

When brainstorming is going well, it is also play. People joke, laugh almost continuously, and suggest outrageous solutions as well as "respectable" ones. As with other forms of play, there is diminished sense of self, an openness to improvisation, and a desire to keep the process going. There is a dynamic give-and-take among group members, and everyone is involved (as in other play, more dominant members of the group will self-handicap to put themselves on equal footing with less dominant members). Time seems to fly by. It doesn't seem like work (appears purposeless), but at the end of the process there are a number of really good ideas.

On the other hand, the results are not always good. Some studies report that brainstorming is no better than asking individuals for ideas. The problem, I think, is that some brainstorming sessions never become playful. Although no one is supposed to criticize, some group members will feel that unspoken criticism is in the air. Or a kind of unspoken hierarchy hampers the freedom of expression. Even very creative groups can fall prey to these problems. I am familiar with one company that is known around the world for its creative output, but which has been vexed by brainstorming sessions that have become stagnant. What I've observed is that group members, all smart people, have internalized the expectation that they will always come up with brilliant ideas, which no one can do. There are not supposed to be judgments made about ideas, but they sense that their peers are judging them. The solution that I proposed was consciously establishing activities that set the group up for play, like playing Twister before the brainstorming actually began. When all these smart people play, the brilliant ideas will flow naturally.

On a larger scale, companies as a whole often talk about promoting creativity and innovation, but they kill their own best ideas and drive off or shackle the people who come up with them. My colleague Lanny Vincent has spent a lot of time and creative energy of his own analyzing why this is. What he finds is that new ideas in companies threaten the existing order. They activate the company's "autoimmune system," which protects the existing corporate structure against products that are "not invented here" or "not sold here." I'm sure there were people in Polaroid who saw the handwriting on the wall and wanted to put resources into research on other products such as digital imaging. The problem is that this takes money away from established corporate divisions who will complain that they are hurting already and that Polaroid will never catch up with companies that already have a market in digital imaging.

In Lanny's view, innovative ideas and the groups who come up with them have to be insulated and protected from the normal corporate structure. New ideas are like newborn babes, weak economically and with little promise of quick returns. I know that in the mid-1990s, if most executives thought about the World Wide Web at all, they considered it to have very little business application. If they knew that employees were using company time to explore the Internet they would consider it a waste of time at best, or grounds for firing at worst.

Lanny finds that ideas have to be nurtured and supported within companies until they can stand on their own. The people who come up with these ideas have to be insulated from executives who would view their activities as a waste of time, counter to corporate policies—and worse, might convince them that this is so. What's more, even when these ideas are protected, they will most likely not survive unless they are midwifed into existence by a company maverick, someone who has a reputation for tolerating and nurturing wild ideas and methods, but who also has a track record of sound business decisions. Sometimes these mavericks are company founders who have turned day-to-day operations over to more established business leaders. I liken mavericks' psychological role in the evolution of a company to the role of transitional objects in child development. Children take their teddy bear or blanket into new situations as a safe and familiar bridge between dependency on parents and a self-sufficient childhood. The mavericks help the corporate establishment feel comfortable in the foreign world that the innovators inhabit, and help the innovators feel comfortable in the corporate world. Yet they really belong to neither.

On an individual level, your creativity also needs to be protected, not only from outside critics, but also from your own internal critic.

Allow yourself to be abundant in your creativity, at first not making judgments about what you think, feel, or do. Simply play with your ideas, with how you do things. When you are stuck, try imagining fifty "impossible" solutions and then let yourself throw out forty-five. One particularly famous scientist I know told me that the secret to his brilliant ideas is that he has a really big wastebasket. He let himself enjoy thinking up and throwing out one hundred bad ideas before finding the single good one.

mastery

There is a great deal of evidence that the road to mastery of any subject is guided by play. Learning a subject by rote can take one only so far. To become a master, the pupil has to go beyond what is known, has to learn what has not been shown by others in the field. Those who study the history of the arts and sciences have many examples of discoveries that came about not through the progression of a planned series of experiments (or at least not a series of experiments that went as planned).

Roger Guillemin, in his long quest that resulted in a Nobel Prize, had a hunch that the accepted and "proven" pituitary control of metabolism, stress, reproduction, and other regulatory hormone-driven physiology was not the whole story. He didn't think it explained the hormonal feedback and regulatory process sufficiently. He was doing microdissections of the pituitary, and as he pondered the possibilities he discovered that the veins draining blood out of the brain and into the pituitary were much larger in capacity than was needed to drain the arterial pituitary flow into that part of the brain. Hm, that's funny, he thought. He soon figured out that the extra volume was from a superhormonal control of the pituitary coming from the brain (the hypothalamus, to be exact).

Most often, new discoveries and new learning come when one is open to serendipity, when one welcomes novelties and anomalies, and then tries to incorporate those outlying results into the broader field of knowledge. As Isaac Asimov said, "The most exciting phrase to hear in science, the one that heralds new discoveries, is not 'Eureka!' but 'That's funny . . .' "

The state that most promotes those serendipitous moments and makes us open to anomalies is one of play. We are most likely to say "That's funny . . ." if in fact we are open to appreciating something funny or unexpected. Otherwise we see only a very unfunny failed experiment. In 1856, an eighteen-year-old William Henry Perkin tried to synthesize the antimalaria drug quinine from a petroleum derivative but failed, ending up with a tarry black mess. The episode could have ended there, but being interested in painting and photography, he saw that when a little of this mess was thinned with alcohol, it would stain cloth bright purple. All cloth dyes at the time were made of natural derivatives that were expensive and often not colorfast. Purple was the rarest and most expensive dye of all. Perkin's aniline dye was the first purple chemical dye, which ushered in a fad for purple clothing that made the 1890s the "mauve decade."

Finally, and perhaps most important, work that is devoid of play is either boring or a grind. We can get pretty far through sheer will-power, and some people have prodigious powers of perfectionism, self-denial, and suffering. Ultimately, though, people cannot succeed in rising to the highest levels of their field if they don't enjoy what they are doing, if they don't make time for play. Having a fierce dedication to grinding out the work is often not enough. Without some sense of fun or play, people usually can't make themselves stick to any discipline long enough to master it.

People always say that you can reach the top by "keeping your nose to the grindstone," but as sports performance specialist Chuck Hogan observes, this is not true. People reach the highest levels of a discipline because they are driven by love, by fun, by play. "The great performers perform as they do, and do so with such grace, because they love what they are doing," Hogan observes. "It's not work. It's play."

Tiger Woods hits thousands of golf shots because he loves it, and he loves it because he plays at it, not works at it. Woods told Ed Bradley of *60 Minutes* that as a kid he would throw balls into the trees, have them drop randomly in the thick rough, and try to make par anyway because it was more fun. People I know at Stanford University, where Woods was an undergraduate, told me that he would hit shots with a slice so extreme that the balls would go over the apartments on the left side of the Stanford driving range and then curve back onto the grass. He did this just for fun, because "sometimes hitting regular golf shots is boring." On a shoot for a Nike commercial, Woods passed the time while the cameras and lights were set up by bouncing his ball on the end of his nine iron forty or fifty times and then whacking it away without ever letting it touch the ground. The director asked if Woods could do this on-camera and it became a hit commercial. "I enjoy creating," he told Bradley. "I enjoy creating shots."

Athletes may not love every minute of training, nor every moment of a game or competition. Sometimes the joy comes from the fantasy of the win. Every athlete I have ever met often feels that they just don't want to start that workout. But when they do, the reason they love what they are doing comes back to them pretty quickly.

The work we do has to be the same way. Part of virtually any job has the possibility to be as enjoyable, as enthralling and creative as when we were kids building sand castles on the beach or flying a kite we created out of sticks, newspaper, and string. The joy has to find its way to us, and us to it.

losing it

If play is so necessary to our work, why do we lose it? The answer is that we are both pushed and pulled away from play. As we get serious about career, get married, have a family, move up the ladder at work, take care of parents, take part in community and religious duties, and work out to stay in shape and prevent health problems, we feel we are inexorably pulled away from any time for personal play. Barbara Brannen, the HR executive I wrote about at the beginning of this chapter, was in this situation. A crisis of some sort is not uncommon for successful people at midlife, but the age for this midlife meltdown has started coming earlier and earlier. People have started taking about a "thrisis" in the thirties and a "quarter-life crisis" in the twenties. I've also seen the same pattern in younger people, even adolescents, who have a jam-packed schedule of schoolwork, homework, after-school activities, community volunteering, tutoring, and test prep. I haven't heard a catchy name for it yet, but they too suffer the same crisis of the soul that comes from pouring every moment of your time and every ounce of your being into others' expectations.

If anyone goes without play for too long, grinding out the work that is expected of them, they will at some point look at their lives and ask (usually silently and privately), "Is this all there is? Is this what I can expect for the rest of my life?" The acquisition of good grades or a big bonus, if not connected to the heart of life, is dispiriting, even if accolades accrue. For some people this loss of faith will happen at sixteen. For others it will happen at sixty. For consummate players, it does not happen.

In addition to being pulled away from play, we are pushed from play, shamed into rejecting it by a culture that doesn't understand the human need for it and doesn't respect it. As I've said before, play is seen as something that children do, so playing is seen as a childish activity not done in the adult world. The message is that if you are a serious person doing serious work, you should be serious. Seriously.

Most of the time, we have so internalized society's messages about play being a waste of time that we shame ourselves into giving up play. There may be people in our lives who tell us to take it easier, have a little fun, but we just can't allow ourselves to do that. Mark, the restaurant manager, was able to have a fun time and get the job done when he was just a restaurant worker. But as soon as he was promoted to manager, he put himself in a parental role. He felt that he shouldn't be seen to be playing, that this was not what a manager did, regardless of whether or not he could be goofy *and* do his job. Mark felt that it would not be responsible to play.

Sometimes I am taken aback by how strongly people block play, by the outright hostility to play. I was recently at a conference in which I spoke about play, and was generally basking in the warm reception given me by other attendees. One speaker who had a highly successful career creating children's television programs got up after hearing me speak and said, "I can get funding for any project I want to do, but what we really ought to be finding funds for is research on play. This stuff is really important." So it came as a shock to me when I was accosted on my way to get coffee.

"Hey, why do you say that you can learn more about play from getting down on the floor and playing with children than you can from reading books?" The guy asking this was perfectly dressed, carried a computer under one arm, was much larger than me, and stood too close. I explained that I was trying to say that you could do all the book learning you wanted and it would still be hard to understand the emotion of play that you could easily get by experiencing it with kids, who are very pure in their play pursuits.

"Well, then, you are being inaccurate," he told me. "You're being irresponsible telling people to play. What are you, a hippie?"

I felt very uncomfortable. I couldn't be sure he wouldn't punch me or something. Clearly, this was someone for whom free play was a threat, and who likely had been prevented from the joy of childhood free-play himself. I asked him who in the conference he had found compatible with his viewpoints, and he turned on his heel and hustled away. (The conference was a major one, revolving around "serious play.")

Though his aggressiveness was unusual, it was not uncommon. I have experienced too many cold and hostile reactions to play when people listen to a full rendition of its nature and importance, and they slowly realize that they have lived a life deprived of spontaneous play. They are struck by the fact that what love they've had in their lives was conditional and based on their performance. To fully realize this in one sitting as an adult can be overwhelming—too much to bear. The reaction is often an intense (but unconscious) defensiveness, a denial that the fullness of one's life has been wasted. The resulting emotion is usually anger at the deliverer of the message.

Joy is our birthright, and is intrinsic to our essential design. For highly competitive, serious people to realize that they have missed this joy can be devastating. Perhaps this fact lies behind the culturally supported idea that people who play are superficial, are not living in the real world, are dilettantes or amoral slackers.

If time and circumstances allowed, I would have gently led him through his own play history. I likely would have asked him to begin to look at the sources of his anger, its intensity (inappropriate to the circumstances), and with as much empathy as possible would have pushed open the door to finding out where this anger comes from.

When I see people react this way, they are generally people who don't play, who pride themselves on not giving themselves over to play. They define themselves as being no-nonsense people who have been rewarded by their parents, teachers, and bosses for their hard work and accomplishments within a narrow norm. Perhaps they have also been punished for not working hard enough.

When I begin to lay out the natural history of play and how necessary it is, for many adults with fixed attitudes about play, the idea of its nature and importance to their lives is too much for them to handle. As they begin to emotionally sense what they have been missing in their lives, have been without the deeply satisfying joy play can bring, their defenses come up. They feel that the idea that we should make a priority of play has to be not only rejected, but also decisively squashed.

I've seen this in my own former career in medicine, which is indeed a serious profession. Of course you don't want to be joking around or appear unsympathetic as you examine and diagnose a patient with a very serious disease, but there are a number of doctors (like Patch Adams, who was played by Robin Williams in the movie of the same name) who have shown that humor and playfulness can create a healing connection between patient and physician. There's a saying that laughter is the best medicine, and I think that is often true.

Yet most signs of lightheartedness are commonly pounded out of students during the boot camp of medical school, which sometimes seems like a cross between a sleep deprivation experiment and a fraternity hazing. Fortunately, the importance of listening to the patient with empathy has been introduced into contemporary medical school curricula, and studies on sleep deprivation have modified on-call schedules. But as a young doctor in training, I liked the hard work, and found the various courses and ward experiences fascinating. It wasn't the cultural norm then that medical students could both be serious and be playful. As a result, I became a covert player. A fellow student and I had to write a demanding thesis and yet we wanted to get in shape and spend time outside, so we decided to measure insulin uptake and a few other physiologic changes that might occur during endurance conditioning. We would draw each other's blood before a great run, and then head out for our fun. Sweaty and happy, we would draw a second sample at the run's end. We had a ball, and the results of the thesis were respectable, though it didn't win any prizes. (Had we been a bit more enterprising, we would have had fun *and* won a few prizes.)

People in offices all over the world also have to take a similarly subversive approach to enjoying their work. Enjoying your work too much might be taken as a sign that you are not working hard enough or don't have enough to do. It might be seen as disrespectful toward people who are overloaded with work (or for those who at least feel overloaded). Happy eagerness about the work might lead others to think you are sucking up to your superiors, are a Pollyanna, or are independently wealthy and don't care if you lose your job. All in all, it's best to take on an energetic but stoic attitude toward the work. Even so, a certain amount of resignation or even cynicism about the workload can take hold.

This is nonsense. We *can* enjoy our work. We *can* have fun. We can discover how to find as much joy as we do when involved in any project, as much joy as we did when we were a kid making paper airplanes and flying them from the roof.

When we lack that feeling of lightness in what we do it should be taken as a warning sign. It should be as alarming as chest pain or shortness of breath when we climb stairs, a high blood sugar reading, or anemia. If we had a simple test for play like we do for diabetes or high blood pressure, we could look at a number and realize that we are in danger. But we don't have such a test. Instead we have a smoldering, play-deficient sense that something is missing in life, that we are not getting the feeling of joy and energy that we once did.

The question is, how do we get that feeling back?

getting it back

There is no single simple recipe for bringing back a sense of play in your life and work. I can't tell everyone to play two games of Scrabble, visit Venice, or try to wiggle your nose, although I suppose that wouldn't be a bad start. There are ways to jump-start play. As I emphasized earlier, I find that physical activity—*movement* of any sort—has a way of getting past our mental defenses. National Institute for Play adviser Frank Forencich, who has established a play-based organization called Exuberant Animal, suggests that people start with standing on one foot on a wobble board, and go on from there. One of our board members at the institute had his own play epiphany when he found himself alone on the beach and allowed himself to skip.

Many years ago, I conducted a study showing that regular physical activity could help seriously depressed women rise out of their depression. This was a yearlong study that involved a five-day-a-week commitment on the part of women who were depressed and unresponsive to antidepressant medication (or refused to take it). Many seemed stuck in self-doubting questions such as, "Why didn't my marriage work out?" or "Why don't I have more friends?" They were referred to this study by local physicians. The beach in San Diego allowed them an uncluttered two-mile track, where, after medical clearance, they were cajoled into sustaining 80 percent of their maximum cardiac output for forty-five minutes four times weekly. An exercise expert and I monitored pulse rates and pulse recovery times until the participants could assay their own efforts. The first three months were titanic struggles, but then the positive effects of conditioning, exercise, and group solidarity allowed the majority (some had dropped out) to note lessening of their depression and improved overall well-being. For the majority of the participants, the maintenance of endurance conditioning has been necessary to prevent a recurrence of their depression.

Even a short walk can lift the spirits. The body remembers what the mind has forgotten. Body play is the first thing that shows up in evolution. Species capable of play show this in their exaggerated

jumps, twists, and turns, quirky off-balance to gain balance movements, done apparently because they are fun to experience. It's like the baby hippo in a small Serengeti pond doing backflips in the water, or a baby orangutan swinging upside down like a crazed pendulum. Since movement is the first thing that shows up in our own development, it can also be the first step we take back into play.

Playing with pets or children also allows us to get past those same, self-censoring impulses that make it so difficult to allow ourselves to play. I've found that a quick film clip of cats playing does as much to get people to understand what play is as an hour of my lecturing. My daughter once called me to complain that her eight-year-old daughter was in an impossibly foul mood, and nothing she said or did would get her out of it. I suggested she just have her go out in the backyard with the dog and just let them be together for twenty minutes. Ten minutes later, my daughter looked out the window and saw the two of them playing. When my granddaughter came back inside, her mood was completely changed and it remained that way for the rest of the day.

These are all short-term fixes, however. To really regain play in your life you will need to take a journey back into the past to help create avenues for play that work for you in the present. This can be done through a complete play history, or it can be done by simply sitting and remembering (and often visualizing) something you did in the past that gave you the sense of unfettered pleasure, of time suspended, of total involvement, of wanting to do this thing again and again. Remember how that made you feel? Remember and *feel* that emotion and hold on to it, because that is what's going to save you. The memory of that emotion is going to be the life raft that keeps you from drowning. It can be the rope that lifts you out of your play-deficient well.

Your task will be to find activities that allow you to recreate that feeling. This is what Barbara Brannen calls "heart play," the kind of play that speaks to your heart and soul. In her book *The Gift of Play: Why Adult Women Stop Playing and How to Start Again*, she describes how she looked at her life and realized that the best memories of play in her childhood were outdoor activities. Her family lived in the country, where she loved spending full days hiking in the woods, or swimming at a nearby lake. That was her heart play. Her husband was someone who enjoyed indoor activities—reading and listening to music, or board games. Barbara loved her husband and naturally wanted to spend recreational time with him, but she realized that her husband's heart play was never going to be hers. Once she realized that she would need time for her heart play and started acting on that realization, she began to experience true play again. She began to feel an excitement with life that she had forgotten. Play became infused in her work.

Setting out to remember those feelings can be dangerous. It can seriously upend your life. If Barbara's marriage wasn't as strong as it was, her husband might have felt she was pulling away when she went on long hikes by herself. Jason, the jeweler I described at the beginning of the chapter, realized that what he really enjoyed about his craft was working with people to understand what they were looking for, what they needed. He realized that he really wanted to spend more time doing that, and so went back to graduate school to get his Ph.D. in clinical psychology. It wasn't an easy road to take. There was a lot of course work and it cost a lot of money. But in the end he found himself doing what he truly loved.

When people are able to find that sense of play in their work, they become truly powerful figures. It can be transformative. I've spent a little time working on a project with Al Gore, and I'm familiar with his past. He grew up as the well-off child of a United States senator, often living in hotels in Washington, D.C. But his father didn't want him to be spoiled, so he had him work on the farm in Tennessee during the summers. The feeling I got from Gore was that he felt the need to always be responsible and dutiful, so he seldom allowed himself to really cut loose and play—at least not in any public setting. My feelings were confirmed when he was running for president. People called him stiff. They sensed that something was being held back. When he "lost" the presidential election, and after an understandable period of decompressing and after taking stock of his priorities, Gore changed. It was as if he was freed to finally pursue what really fueled his soul. He put together a presentation on global warming that became a movie and a book, *An Inconvenient Truth.* In the movie and in person, Gore's joy and his emotion came out. You didn't get the sense that he was holding back anymore. At a conference I attended, Gore made his global warming presentation, and later, one of the most successful motivational speakers of all time, Tony Robbins, told him, "If you had spoken with that passion back in 2000, I believe you would have decisively beaten that guy."

Remember the feeling of true play, and let that be your guiding star. You don't have to become irresponsible or walk away from your job and your family to find that feeling again. If you make the emotion of play your North Star, you will find a true and successful course through life, one in which work and play are bound together. As James Michener wrote in his autobiography:

> The master in the art of living makes little distinction between his work and his play, his labor and his leisure, his mind and his body, his information and his recreation, his love and his religion. He hardly knows which is which. He simply pursues his vision of excellence at whatever he does, leaving others to decide whether he is working or playing. To him, he's always doing *both.*

Chapter Six

playing together

Gail and Geoff, both forty-six, are full professors at a Boston university. Their original shared interest in art history brought them together fifteen years ago. Their early years together provided a heady brew of intellectual challenge, emotional exploration, and romantic passion. They once described themselves as "best buddies who really like sex." But after ten years of marriage, plus three kids and demanding jobs, their intellectual companionship has dwindled. Even thinking about sexual interaction brings up a lot of "issues." They now describe themselves as burned out and maybe wondering how they got into this downer state of being. They are not sure they even like each other anymore.

One Friday night, however, they find themselves duct-taped to each other hand and foot, squirming like Siamese-twin snakes across the floor with three other faculty couples. They "race" toward a line drawn across the floor at the far end of the large community center commons room, where a play therapist conducts the couples' play shop. The laughter is contagious, raucous, and virtually uncontrolled. Sweaty and exhilarated, still laughing to the point of collapse, they reach the finish line. That night, Gail and Geoff make love for the first time in five months, and awaken in the morning as new friends. The change came because they put themselves in a situation in which they had the opportunity to viscerally connect for the first time in a long time. The silly and absurd shared task let them drop their guard and be fully *with* each other. They were able to play.

In the course of taking play histories, I have interviewed a lot of couples, some troubled and some not. Among the troubled couples, some were able to relight the fires of love and some couldn't. The defining factor among couples who were able to find romance again, and even to find new fields of emotional intimacy previously unexplored, was that they were able to find ways to play together. Those who played together, stayed together. Those who didn't either split or, worse yet, simply *endured* an unhappy and dysfunctional relationship.

What has become clear to me now is how play can become the cornerstone of all personal relationships, from everyday interactions to long-term love. In fact, I would claim that sustained emotional intimacy is impossible without play. This is true not only for married bliss, but for continued vitality in long-term friendships.

in the beginning

To understand why play is at the foundation of all personal relationships we need to go back to the very first love relationship—between mother and child. As I've described before, when a nurturing and loving parent is in a spontaneous encounter with a well-fed, safe infant, both radiate contagious mutual joy. This hardwired, spontaneous response is the cornerstone for the child's feelings of safety and intimacy. On close inspection—and using scientific measurement—we see that this encounter indicates a state of play for child and parent. In order to feel intimacy throughout life, the growing child needs to access this earliest of play states. As we get older, the play state is sculpted by our culture—a child in Borneo may play with a doll made of plant fibers, while an American child plays with computers or talking stuffed animals—but it all comes out of the same foundational play states.

As kids grow older, the play they engage in with friends, siblings, and (in a more complicated way) with parents sets the stage for their adult interactions. One woman whom I interviewed for a play history told me that the way she and a friend played with Barbie dolls was exactly how they were in relationships later in life.

"My fifty-year-old friend and I recently dug out our old Barbie dolls," she said. "It was amazing to hear her talk about the way she played with her Barbie at age nine and the way I did—and then to think about our different lives as adults: while *my* Barbie was always attracting men by being a damsel in distress—showing a bit of leg and cleavage—*her* Barbie was a hipster type, smoking cigarettes and wearing Ken's shirts. So here we are over fifty: me, having been married three times, and she, never having married but always with a guy while maintaining a tough-girl attitude. When we played with dolls, neither of us was big on baby dolls—and here we are with no kids between us. And to think we were laying out our adult scripts through play with this doll at age nine."

Some people find such adult-themed play troubling among nine-year-olds. This was also the rap against the Barbie doll originally when it came out in 1959. Dolls before this had always been baby dolls. To give a child an adult-looking doll with hips and cleavage seemed inappropriate. And yet girls loved it. Girls were grappling with adolescent issues and badly needed to be able to address them in the safe arena of doll play. Like the two women mentioned above, they needed play to express core truths about themselves and check out how they themselves felt about those truths, and to figure out how those truths could work in a relationship (both the current friendship and the future relationships with boys).

play in adult relationships

When my dog Jake came to greet me, he would wag his tail, give a stretchlike play bow, and perform a rapid panting akin to human laughter (chimps and rats have also been shown to perform this panting with vocalization when communicating excitement). These were play signals—invitations to get out there and throw the ball or run.

Humans use play signals, too. When we greet each other, we smile and look at the other person with "soft" eyes—looking directly but not staring. We might also raise the eyebrows or lift the chin quickly in greeting. When we get close we might extend the hand for a shake, open the arms for a hug, or purse the lips for a kiss on the cheek. These are an invitation to the other to mirror our expressions, to engage in a ritual bonding with the promise that we will progress to an emotional bonding. And the spirits of safety and trust are communicated nonverbally.

Social scientists often say these gestures indicate that we are not a threat—the open hand extended for a shake, for instance, demonstrates that we hold no weapon. But a display indicating no threat looks much different. People trying not to look threatening will make no eye contact, will stare at a spot on the wall or some object, trying to look busy and inconspicuous. We see these sorts of displays all the time, in subways and grocery stores. These are not engaging social displays—they don't invite mirroring or reciprocation. They invite nothing at all. These are avoidant behaviors usually based on a fear of what an interaction might bring or fatigue with the effort of constructing a social front. In fundamental sociologic terms, they are innate "stranger anxiety" reactions to someone unfamiliar.

If we lived in a world without play, all public adult interactions would model those of subway sitters and elevator riders. It would be a pretty grim world to live in. What play signals do is invite a safe, emotional connection, if even for an instant. Even in casual interactions, the sincere compliment, the remark about the hot/rainy/freezing/damp weather, a joke or sympathetic observation opens people up emotionally. It transforms a grim, fearful, and lonely world into a lively one.

I experienced this transformative power firsthand recently. I was standing in a long line at the pharmacy, and the person at the front was working out some sort of complicated insurance billing question. The line was not moving, and everybody was irritated. One woman walked up, looked at the line, and wandered away. Shortly after, she came back and took another look at the line. I smiled and told her, "We all come here because we really like hanging out here. It's a great place to spend some time." She chuckled and others in line started chiming in. "Yeah, we reconnect with old friends." "It's a great place to pick up cooking tips." "Let's take a bet on how long it will take the last person in line to make it to the pick-up window." Pretty soon everyone was laughing and joking. As I was leaving, the guy who had been standing behind me said he had never had so much fun in line before.

The thing is, before I came to the pharmacy I myself wasn't having that great a day. I was low-energy and had a mild headache. I could have spent the time in line struggling to be patient and not get upset. But I made a conscious decision to take on a playful attitude. And as I walked back into the parking lot I felt very different from how I had felt when I initially walked away from my car. I felt much better about my day, and I felt good that I had helped others feel better, too. And my headache was gone.

Really making emotional contact with people, inviting an emotional closeness either in a casual situation or long-term relationship, requires that we open ourselves to them. It requires that we not put up defensive walls and that we accept others for who they are. Then we can invite others to engage in play.

Kay Kostopoulos, an actress and a drama teacher at Stanford, starts her beginning students not with any sort of acting exercise, but with observation. She has students pair off about two feet apart and look at each other for three solid minutes. A lot of people find this really uncomfortable. People only look this intently at each other if they are in love or in a stare-down confrontation. It's very personal. But the teacher urges people to get over themselves, to stop thinking about how they look and feel, and instead to think about the other person. How would you draw them? How would you describe their face? What might you guess about their personality from their features? How did they look as a child? Is that a little chicken pox scar? Have they lived a life in the sun? After a minute or so, people begin to lose their self-consciousness and really *see* the other person. Then the students are ready to open up and enter the state of play that is the necessary prerequisite for acting.

There are a variety of play behaviors that allow us to open up safely. Play modulates deep psychological fears and insecurities that threaten emotional closeness. Teasing, as I've noted before, is a common way to probe the boundaries of a relationship and address power issues. In general, men engage in teasing more than women, and the teasing can seem rough to someone who is not used to it. "Are you still driving that old junker?" one guy might ask another. "Yeah, but at least it's paid off," says his friend, "unlike your land yacht of an SUV. Your mileage is, what, two gallons to the mile, right?" If someone gets genuinely upset, the friend can say, "Hey, I was just joking around," or "I was just messing with you." The boundaries for such heckling are normally general cultural norms, but body language during the encounter usually primes the teaser to keep it up, or back off.

Jokes, when they contain unrealistic exaggeration, can allow us to safely address real fears without making them seem like accusations. "Sure, you're my best friend now, but as soon as I lend you the money you're off to Bora-Bora and I may never see you again," one friend might say.

Jokes are the minimally invasive surgery of a relationship: they penetrate to a deep emotional level without leaving an entry wound. A good friend of mine, a gifted retired rheumatologist, recently had cardiac arrest on a tennis court. This left him with some brain damage and general confusion. But what gets through to him are jokes. When other friends and I stood around his bed making jokes, we all laughed and it was clear that he was connecting with us. Humor cuts through clutter in the "higher" centers of the brain, straight to the subcortical, emotional centers. That's why the most reasoned political speech can be cemented into long-term memory by a good joke.

Without the various forms of social play we would find it very hard to live together. Society would either lock up like an overheated engine, or we would have to evolve a rigid, highly organized social structure like that of ants or bees. Play is the lubrication that allows human society to work and individuals to be close to each other.

Which is why play is the most important element in love.

love potion no. 1

Rutgers University researcher Helen Fisher, along with Arthur Aron from the State University of New York, Stony Brook, has put people in love into MRI machines to look at their brain activity. They concur that there are three separate brain systems involved in sexually oriented love: erotic love, romantic love, and attachment.

Erotic love (what is called "lust") is the result of the sex drive. It is very nonspecific—we can feel drawn to anyone who is our "type"—and has evolved to push us to get out there, to find a mate, and assure continuity of our genes. The sex drive is immediate and intense, but not infused with cognitive, seasoned wisdom.

Fisher sees romantic love as being much more specific. In romantic love, we are intensely drawn to one person. Those in love can feel a tremendous amount of energy, as if they are on amphetamines, staying up all night or performing huge amounts of work. When romantic love is mutual, lovers may feel that the world has changed, that colors are brighter and food tastier. Or they may have eyes and ears for nothing but the sights and sounds of the beloved. Lovers have willingly sacrificed their own life for the one they love. Fisher believes romantic love has evolved to get us focused beyond the initial intensity of transient erotic preoccupations to a continuing commitment to one mate.

Fisher's third stage of love is attachment: the companionate comfort and connection that we feel with someone after the fusion-frenzy of lust, the idealization, and intensity of romantic love have faded. She hypothesizes that attachment has evolved to keep us with a partner long enough to raise children. In Fisher's view, these three types of love are independent of each other. We can feel lusty toward one, greatly attached to another, and feel romantic love for yet another, which can be confusing, vexing, and, if acted upon, disruptive (to say the least).

A friend of mine, an average Joe with a "thing" for race cars, got the chance to take a Formula One car for a spin around the Laguna Seca track near where I live. He reported that the power of the car was a little frightening, that it was almost as if he had to hold it back. Just a little too much pressure on the accelerator and the back wheels would easily spin out. Love is a little like that. It is such a powerful force biologically that it can easily spin out of control. In order for love to be successful over time, play has to be a part of each stage. And we need to have had sufficient experience in its variety and benefits to be ready to include it when a hormone-saturated brain and body are part of our package.

Play refreshes and fuels a long-term adult relationship. In a healthy relationship it is like oxygen: pervasive and mostly unnoticed, but essential to intimacy. It refreshes by promoting humor, the enjoyment of novelty, the capacity to share a lighthearted sense of the world's ironies, the enjoyment of mutual storytelling, the capacity to openly divulge imagination and fantasies. When nourished, these playful communications and interactions produce a climate for easy connection and a deepening, more rewarding relationship—true intimacy.

Take play out of the mix and, like a climb in the oxygen-poor "death zone" of Mount Everest, the relationship becomes a survival endurance contest. Without play skills, the repertoire to deal with inevitable stresses is narrowed. Even if loyalty, responsibility, duty, and steadfastness remain, without playfulness there will be insufficient vitality left over to keep the relationship buoyant and satisfying.

We know this at some level. That's why a sense of humor always is rated among the most desired traits in surveys about attractiveness. That's also why dating is designed around play activities. Standard dating activities like dinner conversations, movies, and drives in the countryside modulate the sex drive and give people a ritual space in which to come to know each other. Play also accentuates attraction. It's common knowledge that a little risk (one of the prime elements of play) can stoke love's flame, which is why carnival rides (or in my experience a spin on a motorcycle) are also a popular dating activity. Combining all the elements of play together can be a potent love potion.

Of course, as many a fairy tale attests, love potions sometimes go awry. Fisher tells the story of a male graduate student in a colleague's lab who was in love with a female student he knew. Unfortunately for him, the girl did not return his affections. Through his research, the guy knew that attraction was increased by exciting, novel activities, which boost dopamine levels in the brain. So at a conference in Beijing he decided to put science to work. He had heard that rickshaw rides were exciting, so he invited his love interest to take one with him. She accepted, and the day seemed to go perfectly: as the rickshaw driver took them on a wild ride through the city's backstreets, his date was laughing, shrieking, and grabbing hold of him. After the ride was over, she beamed and thanked him. "That was fantastic! Wasn't that a great ride?" The guy was feeling pretty good until she added, "And wasn't that rickshaw driver cute?"

play as sex symbol

Biologist Geoffrey Miller has proposed that play itself is a sexual trait. He thinks that the products of play—art, drama, sports, music—exist because they are part of the human mating display. The analogy that Miller makes is that the arts and humanities are like the peacock's tail. The large tail may start as a small, bright display of feathers that indicate the male peacock's fitness. If the peacock is sickly or a genetic mutant, it won't be able to create such a bright display. So peacocks with bright displays are more successful at mating. In fact, the size of the display gets to be a sort of fetish. The bigger the display, the more successful the peacock is in passing on his genes, so over generations tails tended to get larger and larger. At some point, dragging around a ten-foot Technicolor tail becomes a handicap and an impediment to everyday survival. Yet sexual selection is so powerful that it trumps natural selection. The huge tail does not have any practical survival purpose, but it looks really good to the hens.

In Miller's scheme, stories and toys like the peacock's tail started as indicators of fitness, as signs of an individual's cleverness or emotional intelligence, but then the process started running away with itself. Eventually we ended up with Picasso and Mick Jagger, both of whom have been tremendously successful at mating, even if their art and music don't actually have biological survival advantage.

The problem I have with this theory is that the arts and humanities, all products of the play drive, do have survival advantage. To view them as essentially useless, as many evolutionary scientists do, is to ignore the ways that play and the humanities help us attune to each other as individuals and as a culture. The arts are indicators of emotional intelligence, but they also *produce* emotional intelligence. They help us grow and adapt. Scientists may be slow to appreciate the actual advantages of the humanities, but as they probe a little deeper they will see that the actors, painters, storytellers, and musicians are not just sex symbols because of a pointless evolutionary genetic fetish. They are attractive because they strongly exhibit the play drive that defines our overall design as a creative species, and that we all must exercise to develop fully. A strong play drive is unspoken evidence of fitness to reproduce.

play and players

I usually say someone is a real player if he or she plays a lot. I mean it as a compliment. But in dating, the term "Player" (I'll capitalize it when used in this way) has negative meaning. Healthy dating involves mutual conversation and an assay of the other person. The kind of person who is called a Player is someone who is entertaining himself or herself in narcissistic, strictly manipulative play (the term is almost always used to refer to men, but women certainly can and do engage in the same sort of behavior). For Players, there is no sense of attunement with the other. He or she is not really looking at the date's life and needs. Narcissistic lovers are intensely entitled, goal-driven, with orgasm, entrapment, guaranteed lifelong dependency, or domination as the goal. In real play, the activity (in this case, dating) is enjoyable in itself and done for its own sake. It overrides consciousness of any goal. When someone is deprived of true play, they are more likely to engage in narcissistic play.

Other objects of desire—great car, expensive restaurant, gold jewelry, a home-based media center—can add to the narcissistic thrill, but there is no emotional opening or true connection. That's why Players are dangerous. They invite emotional openness without reciprocating.

I think it's unfortunate that the term "Player" has come to mean something negative. I wish that we had a lot of words for the different types of play and players, just as (it is said) the Eskimos have for different types of snow.

romance and attachment

Romantic love, that is to say, the "deeply *in love*" form of love, is a super-strong force. The idealization and rapture of romantic love has addictive qualities that are similar to drug addiction. In fact, one of the things Helen Fisher and Arthur Aron found when putting people in love into the imaging machine was that the areas of the brain that lit up were the same as those that light up in people on cocaine.

Again, play's moderating influence is needed. Without play, romantic love naturally tends to drift into territoriality, possessiveness, dominance, or aggression. The emotion of romantic love is to feel totally in sync with the lover, but when lovers go out of sync the fall can be hard. A healthy sense of humor or irony is helpful in keeping little problems from exploding into major ones. Play keeps everything in balance, providing resilience and flexibility in a relationship, and allowing couples to rebound from misunderstandings or unrealistic expectations.

There is a definite downside to love. While being in love is intensely pleasurable, it can also be so intense that it is painful. Which is why the metaphors for love are sometimes violent. Cupid shoots an arrow through our heart. We are "love struck." Even the word "passion" comes from the Latin word for "suffer." The violence and suffering are especially intense if the lover is away, or worse yet, doesn't return one's affections. Studies have shown that being lovesick can cause actual physical sickness. Dying of a broken heart is a metaphor, but people are actually more vulnerable to infection, stroke, heart attack, and other ailments while suffering the chronic stress of lovesickness. On the other hand, the evidence that play *increases* immune strength is becoming more and more substantiated.

By keeping romantic love playful, some of that vulnerability to pain or loss is avoided. Play integrates our emotions and behaviors into our larger world. Lovers can be so involved that they shut out the rest of the world. For them, there is nothing but each other, the essential dyad. Whereas play for individuals leads to loss of self-consciousness, when a couple plays, their playfulness can bring the pair back into alignment with the outside world. Their inner feelings and outer selves gain harmony. In neurophysiological terms, their interior emotional regulatory capacities parallel outer-world realities. Play keeps them from losing friends or jobs.

As everyone knows, the most passionate romantic feelings eventually fade. If play has been a part of the relationship from the beginning, less intense sexual attraction and romantic love will remain, joined by the attachment that is the product of long-term emotional closeness.

the coolidge effect

One "problem" is that we are designed to desire novelty. There comes a time when it feels like you have heard every one of your mate's stories at least once and can anticipate what he or she will say next. The amorous boost provided by novelty is called the "Coolidge effect," after President Calvin Coolidge. The story goes that Coolidge and his wife, Grace, were visiting a poultry farm when the farmer pointed out a virile rooster and proudly explained that this bird could mate dozens of times every day. Even though her husband was within hearing range, Grace said, "Why don't you tell that to Mr. Coolidge?"

Without missing a beat, Coolidge asked the farmer, "Always with the same hen?"

"Oh, no," said the farmer, "with a different hen every time."

"Why don't you tell *that* to Mrs. Coolidge?" he said. (There are other versions of this story, but I find this one the most pleasing because it displays a nicely ironic, playful exchange.)

In animals, the Coolidge effect is a well-documented increase in sexual interest when new mates are introduced. Humans, however, don't have to change partners to increase romantic interest. Other novelties will do. Aron confirmed this in his personal relationships lab. Through the years, he has recruited dozens of couples, measured their baseline sense of happiness with their relationship, and then assigned them to do various activities. In one study, one group was ordered to take the popular advice and spend time in activities that were familiar and enjoyable. The control group was told to do nothing different. The third group was asked to find something new to do together, an activity that they didn't normally do.

What Aron found after ten weeks was that the couples that made a point of doing things that were new and unfamiliar had a much higher satisfaction measure than the couples who spent time doing familiar things. Aron hypothesizes that being engaged in novel activities kicks up the the brain's level of dopamine, a neurotransmitter essential for pleasure. In short, their brains are achieving a state of play. The couples who are doing the familiar and pleasurable are also feeling good, but they likely are not *playing*.

Stepping out of a normal routine, finding novelty, being open to serendipity, enjoying the unexpected, embracing a little risk, and finding pleasure in the heightened vividness of life. These are all qualities of a state of play. Those who simply spend more time together don't necessarily start getting along better, particularly if one or both privately consider spending time together a duty. If the couple's basic patterns of interaction don't change, spending time together can even make things worse.

In order to keep things hot, people have to keep growing, keep exploring new territory in themselves and each other. In short, they have to play. A lot of the clinical psychological approach is to get deeper into ourselves by talking about what's going wrong and working on or analyzing the relationship. Although I think it is important to communicate and to be a positive interested listener, healthy, vibrant relationships cannot just be about problem solving. Psychobabble and intellectual explanations of what might be interfering with mutual joyfulness are not as effective as discovering (or rediscovering) what actually *produces* the joy.

Life has a way of creating challenges to that love. One or both partners may experience a loss of job or income, health problems, loss of status in the community, loss of family members. If we are egocentrically wedded to our outer status—being a doctor, or having a "perfect" appearance—then the adjustment to a change in status will be hard. Inevitably, we lose our youth. Our bodies change. Aches and pains come more often and last longer. In some cases, a major disease like multiple sclerosis or Alzheimer's will completely alter the foundation of the relationship. People need a long established play legacy to deal with these changes. People whose play lives have been vibrant, like my physician friend with post–heart attack brain damage, have buffers against travail and suffer less when major change is thrust upon them. Play produces poise and strength. Consummate players can better meet these changes with grace.

Play allows us to embrace and even sculpt the contours of our fates with an ironic humor and a sense of sharing in our common humanity. The lifelong player remembers this and can *feel* it even in the moments of grief, loss, and suffering. This view of life gives us a strength and courage in the face of the suffering and unfairness of the world. If we can continue to play together we will always be able to find emotional closeness, always be able to find novelty and make discoveries not only about those we love, but also about ourselves.

Chapter Seven

does play have a dark side?

Lawrence was a brilliant young guy. He went to a prestigious college and had a great girlfriend and bright prospects. He also liked to play video games. First he liked them, then he loved them, and then he couldn't live without them.

Lawrence began spending all his time playing online interactive games. His girlfriend got fed up and left him. He moved into a house with a couple other online gamers, and they played night and day. In order to eliminate all sources of light that might reflect off the screen and diminish the game-playing experience, they painted the inside of the house black and kept heavy, dark curtains drawn over all the windows.

Lawrence was addicted. He didn't care about the social, economic, or health consequences of his lifestyle. Lawrence's life and that of his roommates was dominated by online games. Eventually he didn't even want to interact with roommates. He decided to live by himself, and now the only "friends" he has are those he connects with online.

A FEW FIFTEEN-YEAR-OLD KIDS in Milwaukee have a few beers with a homeless person, then start throwing sticks and leaves at him before kicking him to death. "It started out as a game," one of them tells a reporter. "We were all laughing, we thought [one of the boys] was joking. But he wasn't." High schoolers in Florida find a homeless person in the woods and beat him up. The kids are caught because they make a video of the assault, a videotape showing them laughing and joking as they attack. When questioned by the police, they say they were "just playing around."

A SOBER-MINDED FAMILY MAN from Oklahoma City gambles a little in Las Vegas and finds he really likes it. Soon he is a high roller, being comped flights to the city and hotel suites—and losing his house and his business. "When I was playing poker, nothing seemed so important," he says now. "I thought my luck would turn, but it didn't."

EXAMPLES LIKE THESE MAKE people wonder whether play can go bad. Are there times when a healthy sense of play can actually turn on us, when a game can go bad? Can play actually be destructive?

This is a question that sometimes makes people doubtful about the wisdom of being more playful or allowing their kids to play more. I understand the doubts. After all, letting yourself be fully playful can seem scary in itself. You might be afraid of looking silly, or feel guilty about wasting time, or appearing frivolous or immature. There is also the danger that the play will get out of hand, that it will actually become a negative force in your life. For many people, that can become another major reason not to play.

Video game addiction in particular has become a scourge in developed countries around the world. Special clinics have opened in the United States, Europe, and Asia to treat people—adults as well as kids—who compulsively play video and computer games to the point that they experience sleep deprivation, disruption of normal life, and even a loosened grip on reality. Addicts may start shaking or sweating when they see computer terminals and suffer withdrawal symptoms when kept away from them. It is not uncommon for addicts to play for forty-eight hours straight. In South Korea, which has been described as the world's most intense gaming culture, one-eighth of the population between the ages of nine and thirty-nine either is addicted to these games or a has a compulsion that borders on addiction. According to *The Washington Post*, at least ten South Korean gamers died in 2005 from the blood clotting that can occur when people sit in the same position for many hours.

Whether or not play can have a destructive or mean element has been a source of debate in the scientific community. Brian Sutton-Smith is a play theorist who has long felt that play is not an innocent activity. He feels that in play we engage in underhanded and duplicitous behavior, and that sadism and cruelty are just as common in play as they are in the rest of life (or perhaps more so).

Sutton-Smith's recent work *The Ambiguity of Play* makes this case strongly. He documents play that contains imaginative stories about being lost, stealing, killing, running away, and so on. Sutton-

Smith sees this as a dark and threatening "phantasmagoria" that is even scarier than real life. In his mind, the existence of this kind of play proves that play has a negative side.

My feeling is that play, by its nature, has been molded by evolution to create a more optimistic, exploratory view of the world and more harmonious social interactions. What some call the "dark side" of play is actually an assortment of cases in which play is being used to deal with difficult emotions or when people are not really playing at all. Sadism or cruelty as a means of gaining control over another is not play. Driven behavior that has compulsive qualities and that cannot be interrupted is not play.

One of the most common examples people give of negative or destructive play is bullying, probably because we can all remember incidences from our youth when we were playing with friends and were mocked, teased, beaten up, ditched, or humiliated, while our tormentors were laughing and having fun. Worse still are instances in which kids physically assault and really injure other kids (or even adults, like the homeless guy mentioned above) while claiming to be "just playing."

In these sorts of cases, it's easy for me to say that this is not play at all. One of the prime characteristics of play is a desire to keep the activity going. If one of the parties involved in the play is stronger than another, they will automatically self-handicap in order to level the playing field and keep the game going. Self-handicapping is easy to see in animal play, like when the polar bear "bit" at the neck of Hudson the husky. If you want to see many more examples, simply do a YouTube search for "dog and cat playing." The dogs are much bigger and scarier, but they crouch low to the ground to make themselves smaller. If they use their paw to swipe at the cat, they do so gently.

A memory of my own childhood provides an excellent example of handicapping among kids. It is November in Chicago, a typical gray, drizzly, cold, early winter afternoon. I am ten years old. About eight or ten of us neighborhood boys are in the process of organizing a sandlot tackle football game. The two captains, Linny Keith and Dougie Weaver, have been chosen by virtue of their age, size, and ability. Dougie is a neighborhood star. My older brother Bruce and I end up on Linny's team. It is clear that Dougie's team is stronger by a wide margin.

After about fifteen minutes of wrangling over what the local rules are to be, Dougie wins the toss and elects to kick off. I am the smallest, youngest, and weakest member of both teams, allowed to play only because of my big brother's presence and the bravado and confidence he provides. So when Dougie sends a beautiful high kick spiraling toward me, everyone can already tell what the outcome of this first play of the game is going to be.

In the middle of the ball's trajectory, Linny yells at the top of his lungs, "No take!" Play stops immediately and a big argument takes place, with Linny finally winning the argument and adding a new "Three no takes unless we want it" rule to the game. The game goes on.

I don't remember the score, but I do remember that our team lost badly, and that no one on our team was particularly upset about it. We spent more time in the game switching teams and changing the rules than actively playing the game, but there we were—unsupervised, no parents, no referees, wallowing around in cold Chicago mud, keeping the game going for as long as we could.

Adult play is not much different. The competitive urge may make us want to dominate the competition in the short term, but if this happens all the time the game gets boring. That's the reason we have the draft system in professional sports—to keep at least some balance among the teams so that weaker teams have a chance. The natural urge to find balance in play is also the reason that people root for the underdog and against teams that win all the time. No one except the most ardent fans want to see the Yankees or the Patriots win every single championship. In some games, like amateur golf, the handicap is an elaborately calculated and explicit part of the official game.

When someone is domineering, aggressive, or violent, they are not engaged in true play, no matter what they are doing. They can be playing Monopoly, or baseball, or jacks, or just joking around, but if they are trying to hurt someone physically or psychologically, what they are doing does not meet all the criteria of play. People may engage in joking or play behavior while they are doing violence because it eases their conscience and fosters denial of their cruelty. That's part of the obscenity and evil of villains like the Joker in Batman comics and movies—the Joker is worse than ordinary criminals because he so clearly has fun while being violent.

A necessary part of sports and games is winning, but even in competition, there are certain rules that govern the play. There is an agreement that participants be "good sports" who can shake hands and respect each other after the contest is decided. The desire for fair play probably runs very deep in our genes. Marc Bekoff, professor emeritus of ecology and evolutionary biology at the University of Colorado, writes clearly and eloquently about the union of play with emergent morality. His long-term field studies of play in wild penguins and coyotes and his in-depth immersion in philosophy established him as a world expert in discerning the linkages between animal social behavior and the bases of altruism. He also has coauthored a pioneering treatise on play theory, "Mammalian Play: Can Training for the Unexpected Be Fun?" which comes out of his detailed multispecies observations of play in the wild. The legacy of graduate students he has mentored (John Byers among them) further establishes his authority. Bekoff describes play signaling in animals as the basis of social trust, the first socially congruent "icon of belief." (In effect, the animal is saying, "What follows this play-bow will not hurt you.") He also shows that animal play operates to level the playing field and promotes fairness. Thus, for Bekoff, justice begins with animal play and is fostered by healthy human play.

In her books *The Primacy of Movement* and *The Roots of Morality*, Maxine Sheets-Johnstone writes about the origins of "knowing" coming from body movement, with play as a major teacher. She has Ph.D.'s in both dance and philosophy, and in her books weaves a compelling series of arguments for a reunion of body and mind in what she observes is a "cephalocentric," unbalanced world.

Everyone has played with someone who takes the game too seriously, who can't stand to lose, and is in a foul mood long after being beaten. These poor sports are no fun. They ruin the game. It's not even enjoyable to win against spoilsports like this. Poor sports are often narcissistically preoccupied, rather than caught up in the shared emotion and loss of self that team sports, at their best, can provide. They aren't playing at all, really.

Part of the attraction of games is that everyone agrees that it is just a game, that it really doesn't matter in the end. Professional athletes have to work at reminding themselves of this, since their paycheck depends on their performance. It's easy for them to forget that it's "just a game" and start taking the losses and wins personally. At that point, the sport can become a grueling chore. Champion Belgian cyclist Tom Boonen talked about his method for countering this feeling:

> I think winning races has to be a joyful thing. It has to be a hobby and not a profession. You keep in your mind that the day you started (cycling), it wasn't to become a professional cyclist or to make money. You became a cyclist because you liked it and you try to win races as a hobby, for the joy of it and the things you get back from it. If you keep remembering that when you are a professional rider, it makes things a lot easier. To win a race is a big thing, but it's only racing. It has to be something you like, and then everything goes well.

In play, we learn how to deal with life's wins and losses with grace. In the end, we learn to shake hands and let the emotions go, something that is useful in "real" life as well as in games. A poor sport can't do so in either arena.

play addicts

What about addiction to video or computer games? In this case there is no aggression or violence against other players.

There are only games that the player cannot stop playing. Could this be viewed as a pathological form of play?

For most players, it is playful to be a gamer. Most young gamers that I am acquainted with are empowered by their success at video games. They are enthralled, challenged, and improve their gaming skills while playing video games. Most are involved in other activities, or if they are young and still dependent on parents, their access to video games is controlled and other options for fun are available. Such games can even have a positive effect on brain development. A medical journal recently published an article showing that surgical residents who played video games were much more accurate and faster with the arthroscopic tools used in minimally invasive surgery. The Air Force has selected airmen who are proficient in video games to fly the pilotless drones that will one day most likely completely replace piloted fighter planes.

My concerns about play revolving around screens (television, computers, Game Boys, games on cell phones and iPods) are essentially that it can be sedentary and isolate people from real-world, human interactions that are an essential part of psychological health. When someone is gaming or watching a screen, there is no engagement in the natural world, no development of the social nuances that are part of maturation in us as a social species.

The intense visual stimuli that screens provide, along with a captivating narrative, can be very seductive playmates. I've seen kids who are happily playing with blocks on the floor, interacting with each other, negotiating, inventing new story lines, being energetic and talkative. And then the television comes on and play stops. Interaction is no more. The story line is set by the box, and the kids are now merely along for the ride, motionless and mute. Single-player video games are similarly attention hogs and socially isolating.

In the real world, the kind of emotional arousal that these screens and games produce is usually discharged through physical activity. Without this physical discharge, kids can become antsy and unfocused. There have been studies that have shown correlations between attention deficits and excessive game play. It will be interesting to see, though, how video games like the Wii, which combine intense physical activity with screen-based game play, change this equation. It may turn out that these types of games are healthier for the players than sedentary games.

The other concern I have with excessive use of screen-based entertainments is that they neglect a deep human need to interact with the material world: to feel the tug of gravity, to physically move through the dimensions of space and time, to feel the physical resistance of solid objects. Much of our interaction with this world is through the hands. Frank Wilson has spent a career studying the relationship between the hand and the brain. In Wilson's view, the hand and the brain coevolved and are closely dependent on each other. People who lose their hands can of course interact with the environment in other ways, but since the human lineage began millions of years ago, the hand has been far and away the primary tool our body uses for manipulating (a word derived from the Latin word for "hand") the world around us.

The hand and the brain need each other—the hand provides the means for interacting with the world and the brain provides the method. Neurologically, "a hand is always in search of a brain and a brain is in search of a hand," as Wilson likes to say.

Wilson feels, as do I, that the hand and the brain are important not only for each other's function, but that the use of the hands to manipulate three-dimensional objects is an essential part of brain development. All over the world, kids play with blocks, fashion mud pies, push around toys, throw balls, build "forts" and "houses." Normal play, the play that I have shown is constantly fertilizing neural growth and complexity, is packed with examples of hand use.

This is the way it has been for millions of years. It is part of our genetic nature. So when kids involve their hands in work, play, and exploration, they are developing their brains in a manner that is in line with our design, the way that primates all have developed during their long evolutionary trajectory.

There is recent scientific evidence that our brains react differently to three-dimensional objects than they do to the two-dimensional representations on video or computer screens. In one particular study, using a brain imaging technique known as fMRI (functional magnetic resonance imaging), a window allowing direct vision overhead was a part of the experiment. When a *real* hand holding a ball was presented in the window, large areas of the brain's visual and associational circuits were activated. When a *picture* of a hand holding a ball was shown, the visual cortex demonstrated similar arousal but the associational areas were virtually silent. Seems as if we are programmed to "see" more comprehensively in natural settings.

Another major reason for my worry about video game and other screen play is the potential for addiction. Although this is still somewhat controversial, most experts see addiction to computer games as comparable to more obvious addictions like opiate addiction.

In the normal human body, opioids are important molecules. One of the jobs of opioids is to block the circuits that produce pain. People describe an opioid high as creating a deep sense of well-being and comfort, as if all the hard or rough edges had been taken out of the world. In normal life, the brain releases a small amount of such painkillers, called endorphins, when we are under major stress or in severe pain. These endorphins and other neural-signaling molecules are the sources of the so-called runner's high that comes after an intense workout. But there are also natural feedback circuits that regulate how much endorphin we get exposed to. When people take an artificial opioid like heroin, codeine, or Percocet, that regulatory mechanism is short-circuited. The brain gets a huge hit of opioid.

The thing about opioids is that their use doesn't automatically lead to addiction. Studies have demonstrated that most people who are prescribed opioids for pain don't become addicted. They appreciate the release from the discomfort and may even like the high, but they easily stop taking them when their illness is over. People who feel chronic psychic pain as a result of abuse or other causes, however, do often become addicted because the opioids offer relief from this psychic pain, a relief they did not believe was possible.

Play also activates reward circuits because it is a beneficial activity. As with endorphins, there are natural regulatory circuits that limit how much play we will engage in. Adults who are healthy and psychologically well balanced will enjoy playing, but after a while they will grow tired of whatever game they are playing and do something else. People who are using the games to escape some other psychic pain, however, will not stop playing. If they do, their pain and anxiety will come rushing back.

The arousal and pleasure that provide this escape can therefore become addictive, with disastrous physical, social, emotional, and cognitive results. The hard-core late-adolescent gamers I have interviewed express real difficulty in differentiating virtual from real. They experience a different quality of stream of consciousness than a nongamer. The nongamer usually enjoys the pretend aspect of his inner life but can easily differentiate it from concrete reality. Not so for the addicted gamer.

In addition, the kind of play that people are addicted to addresses a deep psychological need. Many of those who are addicted to computer gaming are those who don't feel comfortable meeting life's varied and ambiguous challenges. In life, it's often not clear if you are "winning" or "losing." Gaming offers a very controlled world in which victory and defeat can be clear and unambiguous. Part of the reason for widespread game addiction in Japan and Korea may be that those are societies in which there are intense pressures for young people to be high achievers along a very specific and rigid career path, offering little chance for the young to define their own quest.

What about other forms of play? Howard was someone in the grip of a gambling addiction. He told me that the first time he fully experienced unfettered joy was when he hit his first slot machine jackpot and when he then bought a round of drinks for the crowd. He felt powerful, in control and loved, things that he had never really felt he deserved. He was unable to stop fantasizing about another big win after that first hit and kept gambling to experience those feelings again. The young men (predominantly) who are addicted to video games usually feel anxious about their ability to meet the demands of the adult world. Their social skills have not been honed, they are often temperamentally shy, and they feel expectations for success that they don't think they can meet. Online or video gaming offers a world in which they *can* succeed. They are quickly rewarded for success, and failure is easily reversed.

Gaming addiction doesn't represent a "dark side" to play any more than obesity represents a "dark side" to food. On the whole, three-dimensional physical and social play is a "better" form of play, just as a balanced diet is better than one full of sugar hits. The latter provides more instant gratification, but is damaging in the long term. We need both play and food to remain healthy. When we overdo it, the fault lies not in the play or in the food but in ourselves. In fact, being too obsessed with play can be an important indicator about psychological gaps we face in other aspects of our lives. At one point in my life, in my forties, I did a lot of running and spent too much time preoccupied with putting in the most miles and beating my personal best times. Even with all I knew about play then, I regret to say that I didn't realize that drivenness had overtaken much of my life, both personal and professional.

It took a personal life crisis for me to take stock and wake up to what I had known intellectually was the importance of regular authentic play in my life. By being driven to work and "play" feverishly, overcommitted to profession, family, and physical conditioning, I had lost the life-enhancing rewards true play provides. This is not to say that I did not take refreshing vacations or play exuberantly on occasion. But my overall life was pressured, work-oriented, and consistently requiring more than I could accomplish without losing sleep or shortchanging my private needs for personal play.

Looking back, I see that what I would have done differently is to incorporate play earlier and more consistently in my professional life, and to set clear boundaries about working too hard. I think that I artificially separated work and play in my own mind, which was a major blunder. In the family setting I would have tried to incorporate play more into helping the kids with their homework and brought more lightness into the work that needed to be done around the house. I was so driven to accomplish so much in my professional life, and I carried that attitude into the home. The past twenty-one years of play exploration have been transformational for me personally and provided the direct experience that grounds this book emotionally.

breaking the rules

There are some play theoreticians who believe that play is always a good thing and that there are no downsides to play. I don't view play as *always* sweetness and light. Play can be dangerous. People do get hurt. Rough-and-tumble play is typical among all juvenile mammals, including humans. Part of the rough-and-tumble is that, often, someone goes a little too far. In dogs you might see the hurt dog yelp and snap back. In children, one kid might exclaim, "Hey, that really hurt!"

and perhaps strike out. A breakdown of play into a fight can be seen regularly on most open playgrounds. In a healthy situation, the kids pause for a moment while the hurter lets the hurt know that it was unintentional (perhaps nonverbally, by taking the return hit or displaying a concerned expression), and the play resumes. Both have learned something about how far they can go. It might seem paradoxical, but such episodes allow the kids to be closer and even more free in their play, especially now that they have discovered where their own boundaries lie.

When I talk about rough-and-tumble play, most people envision boys, and it's true that boys engage in physical play more often than girls. But girls often engage in a version of rough-and-tumble play that has a strong psychological component. It might involve more role playing, gossip, teasing, or exclusionary clique-formation. A "mean" girl who operates by psychological intimidation and exclusion is the equivalent of a boy bully, both of which interrupt the flow of play. As with physical rough-and-tumble play, kids are hurt. But in a healthy situation, girls learn what constitutes going too far and are closer as a result.

In both cases, I think that we adults are too quick to step in to stop such play. We see the potential for small hurts, hear the squeals and grunts that sound to us like loss of control, and we force the wrestlers to stop. We feel uncomfortable with the gossipy talk and we reflexively step in to make sure that kids are being fair. By doing so, we stop kids from learning on their own and from each other. Of course, when it is clear that there is risk of serious physical injury (more than a bruise or scrape) or permanent psychological damage, we need to act on our responsibilities as parents or playground supervisors. But in most cases it might be far better to let things play out. If things don't go well, we can come back later and ask the kids whether they felt okay about the emotional or physical rough-and-tumble. The kids usually give us the clues. In authentic rough-and-tumble play, the participants squeal, smile, and laugh while hitting, diving, wrestling, chasing, and heckling, and they remain friends after the bout is over.

One of the hopes I have for bringing more healthy play into our future is to have the National Institute for Play help provide preschool teachers with sound, useful information about early rough-and-tumble play. If rough-and-tumble play is squelched because it is seen as chaotic, loud, out of control, its benefits will not be acquired. Kids need a certain amount of this play so that the later stages of development will proceed more smoothly. On the other hand, bullying and exclusionary over-the-top behavior is forestalled if it is nipped in the bud in preschool situations. Teachers and parents need to understand what's normal and what's not.

Many people don't realize that rough-and-tumble play can continue into adulthood, and may be necessary to healthy relationships. We don't physically tackle or punch each other anymore, but we engage in the mock-wrestling we call teasing, joshing, kidding, or joking. Teasing varies by culture and individual temperament, but some form exists everywhere, especially when people are emotionally close. As I mentioned in the last chapter, teasing allows people to go to the edge and just beyond, saying things that may or may not be hurtful if said straight out, offering all parties an escape if they have gone too far. Such teasing is a learned-through-play social skill, with culturally understood boundaries. If the intent is to enlighten or just have fun, teasing and joke-making are great elements of social bonding. If the underlying motive is to put down or humiliate the recipient, it's not healthy.

In many of their movies, Katharine Hepburn and Spencer Tracy used teasing banter to safely explore issues of power in relationships. There's a nice example of affectionate teasing in the movie *Good Will Hunting*, when friends Will (Matt Damon) and Chuckie (Ben Affleck) express their feelings without getting all squishy about it.

CHUCKIE: Hey, asshole.
WILL: What, bitch?
CHUCKIE: Happy birthday.

My family is a collection of champion teasers. At age sixty-two, my cousin Al is in great shape and still plays and coaches hockey. But when we meet, we start verbally cutting each other.

ME: You've lost all your hair.
AL: Well, that's an impressive paunch you've got.
ME: You look like you have hit ninety.
AL: I'm surprised you are still alive.

Occasionally it's a little uncomfortable, but you learn to roll with it, you learn to be ready for it. And underneath it all, I know that it comes from a foundation of love. Teasing about how we look is actually a way of saying, "You look pretty good for your age." I wouldn't tease about how bad he looked if he were on his deathbed. Or maybe I might, if he expected me to tease him, and by not teasing

I was making him worried that he actually looked pretty bad. That's the complex nature of teasing. It's very complexity signals that the teasers are emotionally close, that they know each other well enough to know how far to push it.

Play, by its very nature, is a little anarchic. It is about stepping outside of normal life and breaking normal patterns. It is about bending rules of thought, action, and behavior. Some people use this quality of play as cover for sadistic or cruel treatment of others. "Hey," they might say if others object, "you can't take a little playful hassling? What's wrong with you?" This is not a dark side of play, because it is not play. It's an attack under a false flag. It is an attempt to dominate, demean, or control while hiding behind the bulwark of our cultural assumptions about play being nonthreatening.

In normal play we may still hurt each other by going too far in bending the social rules. Kids may use play to safely explore their capacities for power but end up dominating others. Adults may joke about something that's a little too personal. But when our interactions are based on a foundation of caring, these hurts are corrected and avoided in the future. Bending rules and pushing through limits should happen within the realm of play. They aren't the dark side of play—they are the essence of play.

Chapter Eight

a world at play

I have a wonderful memory: I am on the Serengeti Plain, watching a pride of lions. They are mostly belly up, sleeping and yawning after a big feed. Suddenly, two adolescent lionesses who have been wrestling and rolling around with each other begin a wild ballet. At first, it looks like a fight, but then I see that it is a full-blown, rough-and-tumble dance, choreographed intrinsically by play. It is rhythmic, gorgeous, dominated by curvilinear movements and rat-a-tat paw slaps. There are no signals of aggression. The cats make "soft" eye contact, their hair is smooth instead of bristling, their claws are retracted and their fangs are covered. They make sounds—low shrieks of joy—that are particular to this, and only this, behavior. I almost need a slo-mo camera to catch the intricacy of the moment. And I feel something deep inside me. A visceral thrill, something pure and primal. My linear thoughts get overridden by the epiphany of this moment. It seems as if a spirit of *divinity* has infused these magnificent cats. A spirit of joyousness in physical form. Something more than reflex, something intrinsically creative. I am reminded of Hemingway's *The Old Man and the Sea*, when the title character is at the limits of his endurance in his struggle with a giant marlin. All of Santiago's dreams of storms, fish, women, and fights fall away, leaving only a dream of lions playing on the beach, like cats in the dusk. That is the essential nature of play. It remains when the importance of so much else has fallen away.

A few miles and a million years from the spot where I watch the lions I can find the reason. The great plains of Africa are also the birthplace of modern humans. What we know from the fossil record, molecular biology, and animal behavior studies suggests an interesting story: As the rain forest dried out and became savanna, prehominids descended from the trees and began walking upright. By messing with stones and sticks, and by playing together as juveniles under the protection of adults, the hominids that survived the harsh climate changes on the Serengeti gave rise to animals that were more dexterous with their hands and had better color vision and an upright stance. But it is what was going on inside the brain that was even more important. Our primate predecessors began to imagine and to vocalize thoughts, first in playful gestures, then in the hunt for prey and the search for edible carcasses or nourishing vegetation. Over time, as a result of a lengthening childhood play period, these primates learned to control their tendencies to dominate and fight, and began to be able to reconcile and care for each other. This is how they gained wisdom and survival skills. Those who missed out on play couldn't tell friend from foe, misread gestures, got rejected by potential mates, and floundered. The nonplayers didn't survive. Those who played survived, adapted, and developed skills and capabilities that their ancestors could never imagine.

Which is why I find *hope* in a world in which there is warfare, suffering, vengeance, poverty, and cataclysm. Play has always been a key to adaptation and survival, and I believe it will remain so in the future. One network newsman thinks it is likely that through play we will find a way to change our behavior and better address global warming. In fact, nations may well rise or fall on the basis of their ability to honor our evolutionary prerogative to play.

Why do I say this? Three reasons: social, economic, and personal.

Play sets the stage for cooperative socialization. It nourishes the roots of trust, empathy, caring, and sharing. When we see another human in distress, that distress become ours. Games, sports, and free play between kids set the foundation for our understanding of fairness and justice. Just as my friends and I did in the streets of Chicago, kids argue about the rules of fair play and negotiate about what rules are right for the pick-up game at hand. When sports and games are played as they should be played, organized for the fun of it, kids learn that cheating is wrong and that playing the game the best you can is the thing that matters (although many coaches would rather operate under the Vince Lombardi dictum: "Winning isn't everything, it's the only thing").

Play lowers the level of violence in a society and increases communication. For example, even when there are big social or economic gaps between people, they can often find common ground talking about the local sports teams. If diverse, clashing ethnic or cultural groups can be coaxed to actually play together, the positive effects can be dramatic.

Nate Jones, whom we met earlier in this book, is a Long Beach master mechanic sought out by premier race-car drivers worldwide. He has a heart for (and has established a foundation for) marginalized kids, and uses play techniques to get them to open up and cooperate. He was once invited to a Los Angeles reform school for serious juvenile offenders. Lacking any verbal entrée to them, he packed in a soapbox derby racer he had designed and began to assemble it in the presence of a pack of kids clustered in separate groups, each of which had their own standoffish, too-cool-for-school attitude. Some of the kids started watching what he was doing. As the sleek contraption emerged before their eyes, they wandered over and asked what he was doing. Pretty soon they were helping. Finally, the car was done. Nate picked the littlest kid, who was Hispanic, to sit in the driver's seat, and a midsized but speedy-looking Caucasian kid and an African-American kid as pushers.

They had a ball. They really got into jamming the car down the mild grade at breakneck speed. The guards were dumbfounded. Afterward, they told Nate that they couldn't believe he had gotten the kids to cooperate. Usually, they said, the different racial groups don't even talk to one another. The only time they interacted was to fight. This experience led Nate to bring other cars for the kids to assemble and race. The essentials of play, as with the polar bear and the husky, overrode more hostile and territorial dimensions, and changed a very tense and fragile juvenile prison dynamic. One kid, due to be released, even asked to stay beyond his release date so he could compete in the next car assembly and race.

In the adult world, play continues to be woven into the fabric of our culture. In large part, play *is* our culture, in the form of music, drama, novels, dances, celebrations, and festivals. Play shows us our common humanity. It shows us how we can be free within the societal structures that allow us to live with others. It is the genesis of innovation, and allows us to deal with an ever-changing world.

Economically, developed nations in which people are not merely surviving will rise or fall depending on how well they understand and institute play. That's because what has been called the knowledge economy is being overtaken by the creative economy. In the early days of the twentieth century, industries didn't want workers who could think. They wanted people who could be relied on to repeat the same assembly-line motions efficiently. As other nations gained the ability to host those factories, the "industrialized" countries realized that if they wanted to maintain their standards of living, they would have to work harder or work smarter. Since part of a high standard of living is not working twelve-hour days for six days a week, most people would like to work smarter. A knowledge economy is based on the advantages of its strong educational system, computing power, and analytical abilities. But guess what? It turns out that many of the developing countries are generating a flood of pretty smart people, too. Production design, software coding, computer processor manufacturing, and market analysis can be shipped off to countries overseas.

The advantage that countries like the United States, Britain, France, Germany, the Scandinavian nations, and Japan retain is the ability to invent—to dream up solutions to problems that people may not yet even know they have. Nations that remain economically strong are those that can create intellectual property—and the ability to innovate largely comes out of an ability to play.

As part of my play activities, I have been consulting with high-tech companies over the past few years to help them become more innovative through play. One of them has R&D and other creative engineering labs overseas. I recently spoke with the executive of a worldwide engineering company with labs in the United States, the Czech Republic, and China. He was troubled that the highly trained engineering personnel in China were not coming up with many new ideas, techniques, or technology. The U.S. and Czech teams were doing well, with the United States in the lead. As a result of being convinced that play and innovation were inseparable, he established a "play week" on an island off the China coast and a similar "camp" in the Czech Republic. The Chinese engineers showed a bump in morale and productivity, and thereafter games and free time for imaginative invention were integrated into the workweek. The engineers not only worked better together, but also came up with more effective ways to work and more original solutions for design problems. The interesting thing is that the Czech engineers didn't really take to the extracurricular activities. They already had recreational activities that they enjoyed, and any group activities outside work only took away from those.

the good life

The world needs play because it enables each person to live a good life. What do I mean by that? What do we need to make life meaningful, fulfilling, and worthwhile? This is a question, perhaps *the* question, that has dominated the thinking of philosophers, artists, religious leaders, and common folk through the ages. Their advice, rules, prescriptions, and proscriptions provide guidance on what to eat, wear, say, think, and worship. Even how to die.

Now I will give you my take on it.

My advice on how to live a good life is not so specific. I won't tell you exactly what you can or cannot do, think, or feel. For me, a fulfilling life is one in which we live and grow in accord with our true, core selves, in harmony with our world. A successful life is one in which we are able to fulfill our own basic needs and give of ourselves to others. We are happy when we can live an expansive life, one in which we are aware that we are actively participating in something greater than ourselves—a part of a loving couple, a friendship, a family, an intellectual, social, or spiritual community.

When I take a long-range view of life, through the lens of biology, evolution, psychology, personal experience, and the views expressed by sages through the ages, I see that being playful has an important role in every sphere of our lives. As I've shown in the previous chapters, we are designed by nature to grow and develop in large part through play. Of course we need food, shelter, sleep, and love, but even if we have taken care of simply surviving and reproducing, play is what allows us to attain a higher level of existence, new levels of mastery, imagination, and culture.

When we get play *right*, all areas of our lives go better. When we ignore play, we start having problems. When someone doesn't keep an element of play in their life, their core being will not be light. Play gives us the irony to deal with paradox, ambiguity, and fatalism. Without that, we are like the Woody Allen character in *Annie Hall*, who says, "What's the use? The sun's going to blow up in five billion years anyway."

Living a life of play doesn't mean always choosing the most pleasurable or fun path, however. Joseph Campbell, the brilliant scholar who documented how people across all cultures and all times are essentially living by common mythologies, is probably most famous for his advice to "follow your bliss," but he had to add a clarification because some people took this to mean that they should forgo anything that was unpleasurable or distasteful. I worked closely with Campbell for several years, spearheading the effort that led to his many PBS series. What he believed was that people should find the path in life that fuels their spirit, that speaks to them on the deepest level. But Campbell also showed that this path is sometimes hard. "If your bliss is just fun and excitement, then you are on the wrong path," he would say. "Sometimes pain is bliss."

A friend of mine had an experience when this really hit home for him. A guide was taking him scuba diving on a cold morning. The boat ride out was jarring as the boat slammed through a choppy sea, and the sky looked stormy. My friend was griping about all these things, expressing his unhappiness with the cold, bumpy, wet day. Finally, the guide said, "You know, Josh, you are never going to have peak experiences if you don't let yourself go through some discomfort." At that point he stopped complaining and started seeing things differently. Yes, the sea was choppy, but there was also an awesome quality to the sea, a kind of grand power in the way it tossed the little boat. In the flat overcast, he could see that there was a beautiful chiaroscuro of light and dark where rays of sun broke through the clouds. And they were doing it—the two of them in the midst of nature, about to breathe below sixty feet and a hundred tons of water, where no man had a right to be. He had a great dive.

For me, some of my best memories are of duck hunting with my father and uncles on late-October predawn mornings in Nebraska along the Platte River. For a twelve-year-old boy given the chore of anchoring duck and goose decoys with bare hands in a river flowing with slush ice, it was excruciating. But the prospect of drinking hot chocolate in the duck blind, listening for the first rustling of wings, and hearing the stories of past hunts and family adventures while waiting for daylight was then, and still is right now, evocative of a grounded sense of happiness. While both my own world and the rest of the world have changed and I no longer enjoy hunting, the emotional residue gives me joy. Just hearing the *honk-honking* of Canada geese as they cavort unthreatened year-round along the Carmel River makes me twelve again for as long as the sounds echo from the valley walls.

One of my regular playtimes now is to ride my bike along a steep, winding road near my home. Growing up in the urban south side of

Chicago, I never imagined that I would, at seventy-six, be pedaling up a mountain road surrounded by redwoods, laurels, live oaks, Monterey pines, and blooming chaparral. So the very setting is novel, and for me, is bliss. But the uphill slog is slow, grinding, physically demanding, as my aching thighs and lungs beg for relief. I am no kid. Yet even the uphill climb, while I am puffing hard and wondering if my brakes are dragging (instead of my aging body), is peculiarly emotionally uplifting. As the top beckons I am suddenly out of the woods; the road opens to a panorama of ocean and woodlands. The light is different on every ride, the push worth it, and on the glide home my spirit is clear, happy, at one with body, nature, spirit.

I have had similar moments emotionally when poking through neuroscience literature, and finding an "aha" such as the realization that REM sleep and play share similar brain stem evolutionary biological patterns. So the ride up Robinson Canyon, though physical, and the glow while reading, though sedentary, are for me close to identical "states" of play-being. That is play bliss.

ONE OF THE HARDEST THINGS to teach kids is how to make it past difficulty or perceived boredom to find the fun. "This hike is boring," says my nine-year-old grandson, who loves his video games. But as we keep at it anyway, he slows down and begins to notice details: a four-leaf clover; a hawk carrying a writhing snake aloft; the sound of a strong breeze in the trees. Nature, with all its novelties and the play emotions stirred by its wonders, gets through to kids if the immersion can be tailored to fit their temperament and natural curiosity. They might be on a bike ride and look at a hill and say, "I don't want to climb that. It looks too hard." What they don't know is that there is a really fast, fun ride down the other side. We've done it before, so we know that the effort uphill is worth the payoff on the other side, but all they know is the difficulty. It's the same way with work—they understand the difficulty, but they don't stick with it long enough to understand how satisfying it is to do something really well. You have to make it through the discomfort to find the fun. True play is even one step beyond this.

Making all of life an act of play occurs when we recognize and accept that there may be some discomfort in play, and that every experience has both pleasure and pain. That is not to say that bliss is suffering. My take is that following your bliss may be difficult, demanding, uncomfortable, tedious at times, but not really suffering. In the end, the good feelings we are left with, like my memories of the bike ride or the duck hunt, are far greater than any difficulty we encountered as we played. Advanced play, the black belt of play, comes when we realize this and act on it. As long as we are acting in concordance with our central truth, then the outcome will be positive.

When we fully internalize this ethos, our work is our play and our play is our work, and as Michener noted, we have a hard time telling the difference between them.

play on

At this point, I hope you have become convinced of the importance of play. Now the only "task" is to bring more play back into your life. I say "back" into your life because almost all of us were full-on players when we were little. You only need to stoke a fire that has always been in you. How do you do that? That's the question I often get from people, and while I hate to lay out *rules* for play, I will provide some guidelines.

1. Take your play history

The primary purpose of the play history is to get us back in touch with the joy that we have all experienced at some point in our lives. Find that joy from the past and you are halfway to learning how to create it again in your present life. It also can be a guide to free-flowing empowerment by identifying natural talents that may be dormant or that may have been bypassed.

What we will review here about taking your play history is provided only as a *starter* for a more complete play history available online. I hope this primer will offer insight and motivation to proceed further. From my experience in interviewing individuals who want more play in their lives, I have discovered that it usually requires at least ninety minutes of concentrated but emotionally open, unhurried time to achieve positive results.

This is not a quiz. It is not a test. The play history is a journey through your past and present. It is a time machine, a screen that will show you things you may never have seen clearly, or remind you of things you have long forgotten. Many people find that it brings up more questions than it gives answers. One of its goals is to create a general mental picture of your play attitudes, and color them with emotion-laden scenes. Your current feelings about people, things, and activities are rooted in the emotions you previously experienced and forgot in the natural amnesia of early life.

Start this exercise by spending some time thinking about what you did as a child that really got you excited, that really gave you joy. Was it reading comic books? Building a tree house? Making stuff with Mom or Dad? Did you like doing this with other people, or in solitary? Or both? Were the things that really fueled you more mental or physical? Try to remember the feeling that you had, and recapture it. As part of this remembrance, if visual images spring to your mind's eye, amplify them, let your associations to them flow. To what or to whom do you attach your unalloyed feelings?

Some people have a hard time remembering what they did for play, and even more have difficulty remembering the activity in enough detail that they can really reexperience the feeling it gave them. It's not easy, but it's worth putting in the time to do so. Understand what your unique play temperament is, and how it has manifested itself as you have matured. Then start to identify what you could do in your current life that might let you re-create that playful feeling. Identify activities that fit with cultural norms and your play personality.

Reject judgmental or skeptical thoughts as this exercise moves along. Inventory the whole of your life, with an eye toward play, and look for ways that accentuate joy. Here are some initial questions:

When have you felt free to do and be what you choose?

Is that a part of your life now? If not, why not?

What do you feel stands in the way of your achieving some times of personal freedom?

Are you now able to feel that what engages you most fully is almost effortless? If not, can you recall when you were able to experience such times? Describe. Imagine settings that allow that sort of engagement.

Search your memory for those times in your life when you have been at your very best. (These are usually authentic play times, and give clues as to where to go for current play experiences.)

What have been the impediments to play in your life?

How and why did some kinds of play disappear from your repertoire?

Have you discovered ways of reinitiating lost play that work for you now in your life?

Are you able to imagine and feel that the things you most desire and enjoy are really the things that you ought to have? Why so, or why not?

How free are you now as you play with your spouse or your family? Or do you treat them as an extension of a dutiful responsibility?

Look at your job and what parts mesh well with the person you are. If your work is not satisfying or you are contemplating a major change in your job, be honest about whether it is the right one for you. Have you been able to imagine yourself functioning more joyfully in another setting? This exploration doesn't require immediate practicality or reality. After all, as a kid, your fantasies and pretend stream of consciousness is what added richness to your mental repertoire. The same mechanisms still can be activated, and your brain will ultimately help shape these imaginative flights of fancy to fit what is actually possible. But that doesn't happen without the emotions associated with a "state" of play.

Here is an example of the activation of play histories, in the lives of two of my friends:

Lloyd is a physician I know who realized that some of his happiest memories were working in the kitchen with his mother. Though a busy family physician, he started baking bread on the side, discovered he really liked doing it, and found himself gravitating to more and more intricate bread recipes. He shared his "hobby" with one of his best friends, Miguel, a local high school teacher, who also liked to bake. They set up a more elaborate stove in his basement, and began to leave freshly baked loaves at friends' doorsteps. The friends were enchanted. Soon the basement became an enterprise after friends raved to the local coffee shops, then restaurants. These businesses began to clamor for bread.

Lloyd and Miguel took leave from their day jobs, bought a truck, and really got into it as a business. Lloyd relates a memorable moment that happened early one morning, when he was driving his bread truck to a delivery for the Pebble Beach Lodge, and en route he pulled up beside an M.D. colleague on his way to make hospital rounds. The totally astonished look on the friend's face still evokes raucous laughter in the telling. This hobby became a franchise, which the founders sold when their play urges were fulfilled. Lloyd then had the leisure to pursue other play-based dreams. Long a closet genetics scholar, he became a university-based, self-taught research scientist who could choose his own agenda. Miguel has pursued his love of bicycle racing and now mentors a stable of young racing hopefuls. Each has gone on to be empowered emotionally, financially, and physically.

2. Expose yourself to play

Every day, everywhere, there are opportunities to find play: throw a tennis ball for a dog; pull string in front of a kitten; browse in a bookstore. Here's an old piece of advice that is trite but true: stop and smell the flowers.

The world is full of humor, irony, joy, and objects available for aesthetic appreciation. The trick is allowing yourself to open up to those influences, to see humor in virtually all situations. People begin to close themselves off to play when they start to feel that they should always be serious, always be productive. (After all, we *are* adults!) Soon they don't even notice the opportunities for levity or simple aesthetic appreciation. Activities that by all rights should be play, like a game of golf, can be treated like a self-improvement program or a chance to get ahead in some way. Simply taking a moment to deeply inhale the air after a rainstorm or kick a pile of leaves can be a private little moment of play. More powerful yet are activities that really pull us into play, like getting down on the floor to play blocks with a child. One Ph.D. I know, Fred Donaldson, did this because as a professor he wanted to understand at what point in their lives his students seemed to have lost their desire for learning. He ultimately found that this zeal for learning still existed in preschool nurseries and "open" kindergarten settings but was squelched soon after. This discovery changed his life. He resigned his professorship and focused his life on understanding play and learning. This led to his many contributions through his "original play" workshops.

His "playshops" vary with the audiences. Before the end of apartheid in South Africa, he got white police officers to engage in play with black kids from Soweto, with the result that mixed-race teams engaged in sports and games together. A radical departure at that time, fueled by play. For more conventional groups, he has devised techniques that slowly bring play signals into focus, allowing a group of strangers to become play partners. Usually "soft" eye contact initiates the exercise, with dance movements, curved hand-arm gestures, and a cascade of playful romping with the sounds of much laughter the outcome. (Remember our descriptions of curvilinear movements in the polar bear–sled dog play, or other depictions of rough-and-tumble play?) Donaldson is a real original talent in bringing "states" of play to nursing homes, schools, and settings where the atmosphere is dismal or cooperation has been lacking. His mentoring of others in "original play" has changed many lives.

3. Give yourself permission to be playful, to be a beginner

Probably the biggest roadblock to play for adults is the worry that they will look silly, undignified, or dumb if they allow themselves to truly play. Or they think that it is irresponsible, immature, and childish to give themselves regularly over to play. Nonsense and silliness come naturally to kids, but they get pounded out by norms that demean "frivolity." This is particularly true for people who have been valued for performance standards set by parents or the educational system, or measured by other cultural norms that are internalized and no longer questioned. If someone has spent his adult life worried about always appearing respectable, competent, and knowledgeable, it can be hard to let go sometimes and become physically and emotionally free. The thing is this: You have to give yourself permission to improvise, to mimic, to take on a long-hidden identity. Let your body respond to lessons learned from nature but long suppressed. You can't be truly open to spontaneity if you don't feel comfortable testing novel ways of expressing yourself, pushed along by the pleasure of the action. Play is *exploration*, which means that you will be going places where you haven't been before.

A friend of mine, Daniel, had been skiing for thirty years and was pretty good at it. But it had gotten a little stale and predictable, and he didn't get the thrill out of it that he used to. Like a lot of people, he decided to give snowboarding a try. Also like a lot of people, the first day out he couldn't go ten yards without catching an edge and slamming down hard on the groomed snow. He remembers being amazed at how much abuse the human spine could take and still keep functioning. Little kids blew past him on their boards. At this point, many others in this situation would give up and go back to skis, but Daniel kept at it. He kept falling, but also kept laughing at how ridiculous he looked, even laughing when he got a beaut of a bruise. By the end of the first day he was getting the hang of it, and by the third day he was pretty good. The thrill and challenge of the slopes was back, too.

"It sucks being a beginner again," he told me. "But unless you are willing to do that, unless you can let yourself feel okay about going through the awkward stage, you can't grow. You'll always be stuck in the past."

4. Fun is your North Star, but you don't always have to head north

As I've pointed out before, looks can be deceiving. People may be playing tennis, but they live or die with each point, and ruminate over bad shots or strategy. They are preoccupied with winning and, if they don't win, will see themselves as "losers." This is a tennis game that is clearly not play. Another person might be sweating and grunting as he digs the foundations for a deck, but the truth is that this sort of home improvement is what he loves to do best. When looking for play that really works for you, the easiest way to find what works is to experience what's fun.

The trick is, of course, that some of the really transforming acts of play aren't purely fun. Camping requires packing for the trip and unpacking afterward. Sailing requires boat maintenance, and sometimes delivers wet and cold episodes on the water. Producing any artistic creation includes moments of frustration. As I've noted, Joseph Campbell expounded on the importance of "following your bliss," but that doesn't mean that you should do only things that produce easy fun. If you do, you will be shortchanging yourself.

5. Be active

One of the quickest ways to jump-start play is to do something physical. Just *move.* Take a walk, do jumping jacks, throw a ball for the dog (a double play boost). Motion is perhaps the most basic form of play. We are designed to start moving when we are in the womb. When a grinning and gleeful infant pulls himself up on his feet you can see in his face the pure pleasure of this little triumph. Remember the sea squirt? It has a brain only when it is moving through time and space, and for us such movement is fundamentally pleasurable. We are *alive* when we are physically moving.

Neuroscience research is showing that the fundamentals of perception, cognition, and movement are very closely connected, and that the circuits for higher functions such as planning and recognizing the consequences of future actions require movement. My study of depressed women who were successfully treated through endurance running demonstrated that for me. This research preceded my focus on play, but in retrospect it upholds the power of movement and play to fill an aching heart. Through running, these women discovered a source of vitality and emotional confidence without a lot of intellectual investigation. The physical play bypassed the cognitive roadblocks and built new neural pathways to happiness.

6. *Free yourself of fear*

Fear and play cannot go together. Take a look at your environment and look at where you are unsafe. Is your job in constant jeopardy? Recognize if your body is tight or tense in certain situations. Does your spouse or partner have a critical attitude so that you are always half expecting to be run down for simply existing? If you fear the boss's retaliation or are a naturally anxious person, discover ways to find safe havens. Liz Goodenough, a National Institute for Play adviser and a professor and expert in children's play, says that developmentally we all need "secret spaces" in which we can be safely alone and give ourselves over to needed fantasies if we are to adapt to a challenging world. Find your own secret space. Find out what it is in your surroundings that prevents a sense of trust and well-being that would allow play to emerge. This is not always a simple or easy task, but unless you examine and change these influences, your life force will be trapped. Finding your play is worth it. The ability to play is there in all of us, and is transformative when it is rediscovered.

7. Nourish your mode of play, and be with people who nourish it, too

Practice play. Understand what type of player you are and find ways to engage in your play. It won't happen automatically. In fact, if you have been out of the habit of playing there will be all sorts of habitual resistances and barriers to keep you from playing. Play is nourishing, but you have to take time out for play, just as you would take time out for a meal. And that doesn't mean doing the play equivalent of fast food. Television sitcoms don't usually count, unless you haven't laughed for a while. A lack of play should be treated like malnutrition—it's a health risk to your body and mind.

Be aware of play killers. Part of nourishing your play is putting yourself in an environment that supports and promotes that play. As I've just noted, this is most obvious if you are in an abusive or fearful situation, either in a relationship or at work. If you find yourself often talking about or taking on someone's troubles, or find yourself in a supportive role in which you listen but cannot actually do much to change the situation, that, too, is a play killer. If you are in a relationship in which your interests and ideas are not taken seriously or appreciated, that, too, is a play killer. If people around you cannot learn to understand your need for play, find people who do.

Find the play that feeds your soul, build an environment where people understand your need, and get out there and make it a priority to stay play-nourished.

at play in the world

Bowen White is a medical doctor, a distinguished pioneer in stress medicine, who through a serendipitous suggestion by his physician friend Patch Adams was transformed. While White was on a mercy mission to Sarajevo after hostilities had ended, Patch had him wear a clown costume when he ministered to wounded and orphaned children. To White's surprise, he found a natural outlet for saying and presenting his medically based humanitarian truths to not only kids in need, but also to other professional audiences. He discovered a play personality within himself that had lain dormant since his acting days in college.

In addition to his humanitarian pursuits, White began to gain a wider professional audience by setting up a staged program for medical audiences. He would start a formal presentation on the physiological effects of stress, and then be "called away" for an emergency, staged by the organizer of the conference. Soon, dressed in an outlandish clown costume, he would unexpectedly burst in on the staid conference as Dr. Jerko ("That's yerko, you dolts"), now totally unrecognizable as the professional physician who had spoken earlier. White has a gift for speaking the truth of the human condition most powerfully when in this guise, and in the process he creates wildly unexpected play scenes that fix the truth of his message in the minds of audience members.

White charges thousands of dollars for this sort of routine at business conferences, but he also continues to perform charity work regularly for disadvantaged kids around the world. Recently, White visited an orphanage and affiliated children's hospital in El Salvador with Patch. This was in many ways a sad place—an underfunded and neglected repository for sick and abandoned kids. White and Patch first did their doctoring, and then they worked their magic, making the children laugh and clap for the two American doctors and a party of clowning nurses with big red noses and wild costumes. The kids didn't speak English, but the clowns didn't need to speak any Spanish. They found that comedy and play are a universal language, accessible to all ages in all cultures. It was clear by the end of the day that all had touched one another's hearts.

At one point in the day's events at the hospital orphanage, White was cradling and kissing an infant. After being asked what motivated him to be there, he looked at me and said something that has always stuck with me.

"I really do this for selfish reasons," he said. "I need this kind of connection, and you don't have to come to San Salvador to experience it. There are play and love, and they connect people at the deepest level. Play allows me to enter this situation. Play doesn't solve all the serious suffering, unfairness, or the problems we see in the world, but when you experience it, particularly with a child, it opens your heart, and then you see what's inside. Play helps you regain the mind of the child, and better deal with the major problems and challenges we all face."

For me, this gets to the heart of the matter. Play is how we are made, how we develop and adjust to change. It can foster innovation and lead to multibillion-dollar fortunes. But in the end the most significant aspect of play is that it allows us to express our joy and connect most deeply with the best in ourselves, and in others. If your life has become barren, play brings it to life again. Yes, as Freud said, life is about love and work. Yet play transcends these, infuses them with liveliness and stills time's arrow. Play is the purest expression of love.

When enough people raise play to the status it deserves in our lives, we will find the world a better place.

Acknowledgments

Play is a hugely complex and controversial subject, and I cannot do justice to, or adequately thank, the pioneers and often isolated scholars of play worldwide whose lifeworks continue to inspire and lead. This book stems primarily from clinical observation, my systematic reviews of many life trajectories, and my efforts to integrate the growing objective evidence for the nature and importance of play behavior as it has evolved and as it presents in its wonderful varieties throughout life. As I review the contributions of those who have guided me, and with apologies to those appreciated but who remain unmentioned, I thank:

The late Howard Rome, M.D., of the Mayo Clinic, who encouraged me to see the human condition and diagnosis of its pathologies in the overall context of each lifetime.

Shervert Frazier, former chair of psychiatry at Baylor College of Medicine and subsequently psychiatrist in chief at Harvard's McLean Hospital. He is an inspirational mentor, who opened the professional doors of opportunity in my studies of homicide and play deprivation, and later provided independent scholarly time for me as a visiting fellow. These gifts allowed the science of play to slowly become my life calling.

acknowledgments

Brian Swimme and Joseph Meeker, whose cosmological, historical, and literary expertise gave credence to the idea that ours is a playful universe, with tragic and comedic dimensions.

Howard Suber, Bill Free, and Rozanne Mack, whose humor, sustaining friendship, and wise support is acknowledged with pleasure.

Primatologist Jane Goodall and editor Mary Smith of the National Geographic Society, who encouraged me to enter the world of animal play, and allowed me the privilege of meeting animal play experts and animals at play in the wild.

Bob Fagen, who generously gave me insight into his prodigious scholarship on animal play and with whom Alaskan brown bear play became real.

Brian Sutton-Smith, encyclopedic play theoretician, whose writings and personal discourses opened my eyes to the ambiguities and complexities of play and sealed my fate as an unabashed play advocate.

Michael Mendizza, founder of Touch the Future Foundation, with whom I first discussed play as a "state" rather than a series of behaviors.

Marc Bekoff, insightful, sensitive, brilliant player and play scholar who has framed playful behavior in animals (and us) as the source of fairness and morality.

David Kennard, producer of the PBS series *The Promise of Play*, whose humor, candor, and no-nonsense approach to play clarified the concepts of play personalities.

Scott Eberle of the Strong National Museum of Play, whose historic and phenomenological approach to play has established a framework for understanding play better.

Lanny Vincent and Ivy Ross, who implemented the effort to bring play as I know it into the corporate world.

The Board of the National Institute for Play, and former Chair Werner

Schaer, for their continuing support of the growing vision for play in the world.

Chris Vaughan, who has skillfully condensed my long essays and distilled our dialogues into what otherwise might have been a thousand-page book.

Agents Howard Yoon and Gail Ross, who persevered in their belief that this was a worthy project.

Lucia Watson, cheerful and clear-minded editor, always encouraging and tolerant.

And with deep gratitude to the approximately six thousand people from all walks of life whom I have interviewed over the last thirty-five years, who offered the details of their play lives and whose anecdotes have become data. In some of the play stories that appear in this book, I have altered names or details to protect the privacy of those who have been so generous and forthcoming.

Carmel Valley, September 7, 2008

Index

abused children, 26, 83
academic competition, 111
Adams, Patch, 148, 216–17
ADHD (attention-deficit/hyperactivity disorder), 99–100
Admiralty Island (Alaska), 27–29
adolescents
  and play, 108–18
  rites of passage in, 118–22
adults, 6–7
  play organized by, 105
  play and relationships, 160–64
Africa, 195–97
aggression, 90
Allen, Woody, 202
Alzheimer's disease, 71
Ambiguity of Play, The (Sutton-Smith), 177–78
amygdala, 33
animals and play, 26–30
  ants, 30
  bison, 30
  cats, 32
  chimpanzees, 55–57
  "Coolidge effect" on, 172
  dogs, 21–24, 52, 55, 57–58, 160
  fish, 30, 50
  grizzly bears, 27–29, 31
  hippopotamuses, 30
  mountain goats, 31
  neoteny, 55–58
  octopuses, 30
  peacocks, 167–68
  play-fighting behavior, 29–30
  polar bears, 21–24
  pouncing behavior, 32
  rats, 32, 38–40, 42, 99–100, 102
  ravens, 29–30
  river otters, 37–38
  sea squirts, 47–48, 73
  self-handicapping behavior, 178–80
  stalking behavior, 23
  wolves, 51–54, 57
Inconvenient Truth, An (Gore), 154
Annie Hall (movie), 202
ants, 30
Aron, Arthur, 164–65, 170, 172–74
artist/creator, 69
artistic expression, 61–62
arts, 168
Asimov, Isaac, 142
assembly-line work, 99
attachment, 165, 170–71
attention-deficit/hyperactivity disorder (ADHD), 99–100
attunement, 81–83

index

authentic play, 104
Avischious, Gary, 115–17
avoidant behaviors, 161

Barbie doll, 159–60
beginner (in play), 211–13
behavior
  avoidant, 161
  juvenile, 48–51, 55–58
  mating displays, 167–68
  play, 163–64
  play bow, 52, 160
  play-fighting, 29–30
  predatory, 23, 32
  self-handicapping, 178–80
  survival, 57
  wolves versus dogs, 52–54
Bekoff, Marc, 181
belonging, 88
Bentham, Jeremy, 42
biology of play, 5, 24–26
birds, 50
bison, 30
body and movement play, 83–85
Boonen, Tom, 182
boredom, 204–5
Bowen, William, 117
boys, 190
Bradley, Ed, 143–44
brain, 33
  amygdala, 33
  brain stem, 42, 61, 204
  cerebellum, 34
  cortex, 39
  dopamine in, 173
  enriched environments for, 38–40
  evolution with hand, 184–86
  frontal cortex, 34
  mapping perceptions in, 35–37
  neural connections, 36, 40–41, 48–49, 58–59
  plasticity of, 55–58
  prefrontal cortex, 33
  REM sleep, 41–42, 204
  river otters, 37–38
  simulation and testing, 34–35
  size of, and play, 33
brain-derived neurotrophic factor (BDNF), 33, 100
brain stem, 42, 61, 204
brainstorming, 137–39
Brannen, Barbara, 123–24, 144, 153
Branson, Richard, 67
bullying, 178
Byers, John, 33–34, 181

Campbell, Joseph, 202, 213
cats, 32
celebratory play, 91
cerebellum, 34
c-Fos genes, 102
children and play, 6, 78–80
  abused children, 26, 83
  attunement, 81–83
  boredom, 204–5
  creative play, 92–94
  friendship and belonging, 88
  imaginative play, 86–87
  in utero, 80–81
  learning and memory, 100–103
  movement play, 83–85
  narrative play, 91–92
  object play, 85–86
  parallel play, 88
  parental overcontrol, 94–97
  "perfect" kids, 111
  ritual play, 91
  rough-and-tumble play, 32, 79, 88–91
  schools, 98–100
  social play, 87–88
  toys, 39–40, 104
chimpanzees, 55–57
China, 200
chordates, 47
Clooney, George, 66
cognition, 34
collector, 68–69
Comedy of Survival, The (Meeker), 114

community life, 62–63
competitor, 67–68
competitor (runner), 59
computer game addicts, 177, 183–89
Connally, John B., 94–95
Coolidge, Calvin, 172
"Coolidge effect," 171–74
cooperative socialization, 197–99
cortex (brain), 39
Costas, Bob, 70
creative play, 92–94
creativity, 127, 134–41
crisis, 144–45
cruelty, 178
Csikszentmihalyi, Mihaly, 17
culture, 159, 199
Czech Republic, 200–201

Damon, Matt, 66
Dangerous Book for Boys, The (Iggulden and Iggulden), 79
dark side of play, 176
  breaking the rules, 189–93
  destructive play, 178
  play addicts, 177, 183–89
  poor sports, 180–82
  self-handicapping behavior, 178–80
Darwin, Charles, 44
dating, 166, 169
daydreams, 93
dementia, 58, 71
depression, 126, 150–51, 214
developed nations, 199–201
development, of brain, 36, 40–41, 48–49
Diamond, Marian, 38–40
director, 68
dogs, 21–24, 52, 55, 57–58, 160. *See also* wolves
dolls, 159–60
Donaldson, Fred, 210–11
dopamine, 173
dorsolateral prefrontal cortex, 33
drives, biological, 42–44

Eberle, Scott, 18–19
Edelman, Gerald, 35–36
Einstein, Albert, 93
El Salvador, 217
emotional immaturity, 58
emotional intelligence, 32, 168
emotions, 20–21, 60–61, 152–55, 162
endorphins, 186
engineers, 16
enriched environments, 38–40
enthusiast (runner), 60
Erikson, Erik, 58
erotic love, 164–65
evolution, 44, 55, 58
exerciser (runner), 59
exploration, 212
explorer, 67
exposure to play, 210–11
Exuberant Animal, 150

Fagen, Bob, 27–29, 31–32
fair play, 181
fantasies, 93
fear, 214–15
Feynman, Richard, 136
fish, 30, 50
Fisher, Helen, 164–65, 167, 170
flow, 17
Forencich, Frank, 150
Freud, Sigmund, 218
friendship, 88
frontal cortex, 34
frontal lobe, 100
Frost, Joe, 89–90
fun, and challenge, 213

Garten, Ina, 68
Gates, Bill, 115
Gellhorn, Martha, 129–30
Gift of Play, The (Brannen), 153
girls
  and dolls, 159–60
  meanness in, 190
golden retriever, 53–54, 57

index

Goleman, Daniel, 32
Goodall, Jane, 67
good life, 201–5
good sports, 180–81
Good Will Hunting (movie), 192
Goodenough, Liz, 214–15
Gordon, Nikki, 100
Gore, Al, 154
grandparents, 98–99
grizzly bears, 27–29, 31
Grove, Andrew, 131–32
Guillemin, Roger, 63–64, 141–42

hand/brain evolution, 184–86
Harcourt, Robert, 49
Harold and the Purple Crayon (Johnson), 64–65
Harvard admissions, 111
Hawaii (Michener), 26
heart disease, 71
heart play, 153
Hemingway, Ernest, 195–96
Hepburn, Katharine, 192
hippopotamuses, 30
hobbies, 61
Hogan, Chuck, 143
hominids, 196–97
Huizinga, Johan, 19–20
humanities, 168

IDEO corporation, 93
imagination, adult, 36–37
imaginative play, 86–87, 131–32
innovation, 134–41, 200
Intel, 131–32
intellectual property, 200
internal critic, 140
Internet, 140
Iwaniuk, Andrew, 33

Jet Propulsion Laboratory (Cal Tech; JPL), 9–11
job, self-evaluation at, 208
joker, 66
jokes, 163–64
Jones, Nate, 10–11, 115, 198–99
joy, 147
juvenile period, 48–51, 55–58

Keillor, Garrison, 70, 92
kinesthete, 66–67
knowledge economy, 199
Kostopoulos, Kay, 162–63

La Doone, Brian, 21–24
Labrador retriever, 53–54
Lauer, John, 66
Lawler, Phil, 112–13
learning, 100–103, 141–44
Leno, Jay, 69
Lombardi, Vince, 198
longevity, 72
love
  attachment, 165, 170–71
  erotic, 164–65
  romantic, 165, 170–71
Luhrmann, Baz, 63
Lynne, Gillian, 12–13, 66, 110

"Mammalian Play" (Bekoff), 181
mammals, 50. *See also* animals and play
mastery, 141–44
mating displays, 167–68
Matisse, Henri, 69
mavericks, 140
medical school, 148–49
Meeker, Joseph, 114
memory, 100–103
Michener, James, 26, 155, 205
Miller, Geoffrey, 167–68
Moore, Gordon, 131–32
mother-child relationship, 158–60
motion, 214
mountain goats, 31
murderers, 26, 89

Naperville School District 203 (Illinois), 112–13

narcissistic play, 169–70
narrative play, 91–92
National Geographic Society, 72
National Institute for Play, 150, 191
Nelson, Josh, 33
neoteny, 55–58
neural connections, 36, 40–41, 48–49, 58–59, 109, 127–28
neuroscience research, 214
New York Times, The, 5, 111
New Yorker, The, 64–65
No Child Left Behind, 99
nongamers, 187
nonplayers, 197
novelty, 171–74

obituaries, 5
object play, 85–86
octopuses, 30
offspring numbers, 49–50
Okinawa, Japan, study, 72
Old Man and the Sea, The (Hemingway), 195–96
opioids, 186
opportunities for play, 210–11
otters (river), 37–38
Outward Bound, 120–21

Panksepp, Jaak, 33, 60–62
panting behavior, 160
parallel play, 88
parents
  joy and pain of, 121–22
  overcontrolling, 94–97
peacocks, 167–68
PE4life, 113
Pelligrini, Anthony, 89
Pellis, Sergio, 33
perceptual experiences, 35–36
"perfect" kids, 111
Perkin, William Henry, 142
Person, Ethel, 37
physical activity, 150–51, 213–14
physical education, 112–13
plasticity, of nervous system, 55–58
play, 3–4
  in adolescents, 108–18
  adult-organized, 105
  in adults, 6–7, 59–65
  in animals, 21–24, 26–30
  authentic, 104
  benefits of, 4–6, 9, 49
  biology of, 5, 24–26
  body and movement, 83–85
  and brain development, 33–42
  celebratory, 91
  in children, 6, 77–122
  costs of, 48–51
  creative, 92–94
  dark side of, 175–93
  defined, 15–21
  drive for, 42–44
  emotions of, 20–21, 60–61, 152–55, 162
  gifts of, 103–8
  and good life, 201–5
  guidelines for, 206–16
  imaginative, 86–87, 131–32
  life without, 6
  narcissistic, 169–70
  narrative, 91–92
  need for, 21–24
  object, 85–86
  in office, 9–11
  properties of, 17–18
  purpose of, 30–33
  and relationships, 157–74
  rough-and-tumble, 32, 79, 88–91, 189–93
  self-organized, 105–8
  six-step process in, 18–19
  social, 62–63, 87–88, 160–64
  transformative-integrative, 92–94
  and universe, 44–45
  and work, 126–55
play bow, 52, 160
play deficit, 43
play-fighting, 90

play history, 26, 63, 206–10
play killers, 215–16
play personality, 12, 65
  artist/creator, 69
  collector, 68–69
  competitor, 67–68
  director, 68
  explorer, 67
  joker, 66
  kinesthete, 66–67
  storyteller, 70
play signals
  animal, 21–24, 52, 160, 181
  human, 25, 160–61
Player, 169–70
polar bears, 21–24
Polaroid, 135, 139
poor sports, 180–82
Positive Intervention Through Play, 93
practical jokers, 66
predator stalking, 23
pre-hominids, 196
Primacy of Movement, The (Sheets-Johnstone), 181
problem-solving skills, 10–11
professional athletes, 182
Promise of Play, The (PBS series), 93
psychic pain, 186–87
puberty, 109

Ratey, John, 112
rats, 32, 38–40, 42, 99–100, 102
ravens, 29–30
rebound play, 43
relationships and play
  adult, 160–64
  "Coolidge effect," 171–74
  mother-child, 158–60
  narcissistic, 169–70
  romantic, 164–67, 170–71
REM sleep, 41–42, 204
reproduction, 49–50, 167–69
rites of passage, 118–22
ritual play, 91
Robbins, Tony, 154
Robinson, Sir Ken, 12–13, 66
Rogers, C. J., 51–54
role play, 100–103
romantic love, 165, 170–71
Roots of Morality, The (Sheets-Johnstone), 181
rough-and-tumble play, 32, 79, 88–91, 189–93
rules, 189–93
runners, 59–61
Runner's World, 59–60

sadism, 178
Salk, Jonas, 63–64
salmon, 50
Sarajevo, 216
schools, 79, 98–100, 111
screen-based entertainment, 177, 183–89
sea squirts, 47–48, 73
secret spaces, 215
self-handicapping behavior, 178–80
self-organized play, 105–8
self-reliance, 120–22
sense of humor, 166
serendipity, 142
Serengeti Plain, 195–96
Sesame Street, 102–3
Sheets-Johnstone, Maxine, 181
simulation, 34–35, 100–103
Siviy, Stephen, 102
60 Minutes, 143–44
Slate, 111
smiling, 25
Smith, Ozzie, 115
social play, 62–63, 87–88, 160–64
socialization, 32
socializer (runner), 60
soft eye contact, 211
South Africa, 211
South Korea, 177
Spark (Ratey), 112
sports, 115–17, 133, 180
Stevens, Dave, 137

storyteller, 70
storytelling, 91–92
stranger anxiety, 161
superhero play, 90
survival behavior, 57
Sutton-Smith, Brian, 177–78

teachers, 98, 191
team-building exercises, 134
teasing, 163, 191–93
Texas Children's Hospital (Baylor College), 24–25
Texas Tower massacre (Austin), 94–97
Thatcher, Margaret, 69
toys, 39–40, 104
Tracy, Spencer, 192
transformative-integrative play, 92–94

United States, 200
Upton, Mindy, 98

video game addiction, 177, 183–89
Vincent, Lanny, 16, 139–40
violence, 198
Virginia Tech massacre, 94
visualization, 93, 114–15

Washington Post, *The*, 177
White, Bowen, 110, 216–17
Whitman, Charles, 94–97
Wii, 184
Wilson, Frank, 184–85
Winerip, Michael, 111
Winfrey, Oprah, 68, 115
winning, 213
wolves, 51–54, 57. *See also* dogs
Woods, Tiger, 143–44
work
  creativity and innovation in, 134–41
  losing play in, 144–50
  and mastery, 141–44
  play at, 126–34
  returning play to, 150–55
  self-evaluation at, 208
  team-building exercises, 134
World Wide Web, 140

Photography Credits

Pages 20, 51, 53: Pamela Schwarz

Page 22 (both): Norbert Rosing

Page 23 (both): © *National Geographic*

Page 38: © Mitsuaki Iwago

Page 52: © Daniel J. Cox/Getty Images

Page 54: Liza Mastrella

Page 56 (bonobo with baby; adult bonobo): © Steve Bloom

Page 57: Courtesy Archives, California Institute of Technology

Page 74: Erik Petersen

Page 82: © Kraig Scarbinsky/Getty Images

Page 151 (orangutan): © *National Geographic*

Page 196: © Jonathan Scott